METHODS IN MOLECULAR BIOLOGY

Series Editor
John M. Walker
School of Life and Medical Sciences
University of Hertfordshire
Hatfield, Hertfordshire, AL10 9AB, UK

For further volumes:
http://www.springer.com/series/7651

Type-1 Diabetes

Methods and Protocols

Edited by

Kathleen M. Gillespie

Diabetes and Metabolism Unit,
School of Clinical Sciences, Southmead Hospital,
University of Bristol, Bristol, UK

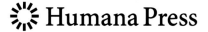 Humana Press

Editor
Kathleen M. Gillespie
Diabetes and Metabolism Unit
School of Clinical Sciences
Southmead Hospital
University of Bristol
Bristol, UK

ISSN 1064-3745   ISSN 1940-6029   (electronic)
Methods in Molecular Biology
ISBN 978-1-4939-3641-0   ISBN 978-1-4939-3643-4   (eBook)
DOI 10.1007/978-1-4939-3643-4

Library of Congress Control Number: 2016942863

Printed on acid-free paper

This Humana Press imprint is published by Springer Nature
The registered company is Springer Science+Business Media LLC New York

Preface

It is now possible with great accuracy to identify individuals at risk of future type 1 diabetes, yet our understanding of the pathogenesis, the rapid increase in incidence, and the best future therapies are reliant on high-quality research utilizing the best possible methodologies. The aim of this book is to present current methodologies to predict and understand the pathogenesis of type 1 diabetes for clinical and non-clinical researchers.

Bristol, UK   *Kathleen M. Gillespie*

Contents

Contributors

RACHEL J. AITKEN • *Diabetes and Metabolism Unit, School of Clinical Sciences, Southmead Hospital, University of Bristol, Bristol, UK*

CLAIRE M. BRADFORD • *Centre for Diabetes and Metabolic Medicine, Blizard Institute, Barts and The London School of Medicine and Dentistry, Queen Mary University of London, London, UK*

MARY N. DANG • *Centre for Diabetes and Metabolic Medicine, Blizard Institute, Barts and The London School of Medicine and Dentistry, Queen Mary University of London, London, UK*

DECIO L. EIZIRIK • *ULB Center for Diabetes Research, Medical Faculty, Universite' Libre de Bruxelles (ULB), Brussels, Belgium*

KAREN T. ELVERS • *Diabetes and Metabolism Unit, School of Clinical Sciences, Southmead Hospital, University of Bristol, Bristol, UK*

KATHLEEN M. GILLESPIE • *Diabetes and Metabolism Unit, School of Clinical Sciences, Southmead Hospital, University of Bristol, Bristol, UK*

YOUJIA HU • *Department of Endocrinology and Metabolism, School of Medicine, Yale University, New Haven, CT, USA*

AIZHAN KOZHAKHMETOVA • *Diabetes and Metabolism Unit, School of Clinical Sciences, Southmead Hospital, University of Bristol, Bristol, UK*

THOMAS KRONEIS • *Institute of Cell Biology, Histology & Embryology, Medical University of Graz, Graz, Austria; Sahlgrenska Cancer Center, University of Gothenburg, Gothenburg, Sweden*

EVY DE LEENHEER • *Institute of Molecular and Experimental Medicine, Cardiff School of Medicine, Cardiff University, Heath Park, UK*

T.J. MCDONALD • *Blood Sciences, Royal Devon and Exeter NHS Foundation Trust, Exeter, UK; University of Exeter Medical School, Exeter, UK*

GEORGINA L. MORTIMER • *Diabetes and Metabolism Unit, School of Clinical Sciences, Southmead Hospital, University of Bristol, Bristol, UK*

JAMES A. PEARSON • *Institute of Molecular and Experimental Medicine, Cardiff School of Medicine, Cardiff University, Heath Park, UK*

JIAN PENG • *Department of Endocrinology and Metabolism, School of Medicine, Yale University, New Haven, CT, USA*

M.H. PERRY • *Blood Sciences, Royal Devon and Exeter NHS Foundation Trust, Exeter, UK*

IZORTZE SANTIN • *ULB Center for Diabetes Research, Medical Faculty, Universite' Libre de Bruxelles (ULB), Brussels, Belgium; Endocrinology and Diabetes Research Group, BioCruces Health Research Institute, CIBERDEM, Spain*

REINALDO S. DOS SANTOS • *ULB Center for Diabetes Research, Medical Faculty, Universite' Libre de Bruxelles (ULB), Brussels, Belgium*

TERRI C. THAYER • *Institute of Molecular and Experimental Medicine, Cardiff School of Medicine, Cardiff University, Heath Park, UK*

LI WEN • *Department of Endocrinology and Metabolism, School of Medicine, Yale University, New Haven, CT, USA*

ABBY WILLCOX • *Diabetes and Metabolism Unit, School of Clinical Sciences, Southmead Hospital, University of Bristol, Bristol, UK*

ALISTAIR J.K. WILLIAMS • *Diabetes and Metabolism Unit, School of Clinical Sciences, Southmead Hospital, University of Bristol, Bristol, UK*

F. SUSAN WONG • *Institute of Molecular and Experimental Medicine, Cardiff School of Medicine, Cardiff University, Heath Park, UK*

REBECCA WYATT • *Diabetes and Metabolism Unit, School of Clinical Sciences, Southmead Hospital, University of Bristol, Bristol, UK*

JODY YE • *Diabetes and Metabolism Unit, School of Clinical Sciences, Southmead Hospital, University of Bristol, Bristol, UK*

LIPING YU • *Barbara Davis Center for Childhood Diabetes, University of Colorado School of Medicine, Aurora, CO, USA*

Methods in Molecular Biology (2016) 1433: 1–9
DOI 10.1007/7651_2015_289
© Springer Science+Business Media New York 2015
Published online: 13 December 2015

Type 1 Diabetes: Current Perspectives

Aizhan Kozhakhmetova and Kathleen M. Gillespie

Abstract

Type 1 diabetes, resulting from the autoimmune destruction of insulin producing islet beta cells is caused by genetic and environmental determinants. Recent studies agree that counterintuitively, the major genetic susceptibility factors are decreasing in frequency as the incidence of the condition increases. This suggests a growing role for environmental determinants but these have been difficult to identify and our understanding of gene/environment effects are limited. Individuals "at risk" can be identified accurately through the presence of multiple islet autoantibodies and current efforts in type 1 diabetes research focus on improved biomarkers and strategies to prevent or reverse the condition through immunotherapy.

Keywords: Type 1 diabetes, Autoimmunity, Insulin, Islet autoantibodies, Genes

1 Introduction

Type 1 diabetes (T1D) represents approximately 10 % of diabetes overall and results from the autoimmune destruction of insulin producing beta cells in the islets of Langerhans [1]. The condition is usually diagnosed on clinical grounds and symptoms appear when approximately 70–80 % of pancreatic islets are destroyed [2]. Individuals with T1D cannot survive without insulin replacement and, even when treated with insulin, remain at risk of complications including nephropathy, retinopathy, and coronary heart disease.

Although commonly associated with onset in childhood and adolescence, with a peak age at diagnosis of 12 years, approximately half of all cases of T1D are diagnosed in adulthood [3]. Epidemiological studies show that the incidence of T1D is unequally distributed in the world's population, with a high incidence rate in Caucasians (40/100,000/year in Finland) and a relatively low rate among Asian and South American populations (0.1/100,000/year) [4]. The incidence of the condition has been increasing rapidly in recent decades for unknown reasons: the current rate of increase is 3 % per year worldwide [5]. If present trends continue, doubling of new cases of type 1 diabetes in European children younger than 5 years is predicted between 2005 and 2020, and prevalent cases younger than 15 years will rise by 70 % [6].

2   Genetic Susceptibility to Type 1 Diabetes

The genetic background of T1D is very complex influenced by combinations of genes and environmental determinants. Identical twins with evidence of autoimmunity studied over a long follow-up period were concordant in over 50 % indicating that the etiology of the condition is approximately half genetic and half environmental [7]. The life-long risk of developing the disease in a child born in a family with T1D is estimated as 20 %, 8 %, 5 %, or 3 %, if the child has two affected first-degree relatives, an affected sibling, father, or mother, respectively [8]. The younger the individual is at diagnosis, the greater the risk to siblings [9, 10].

The importance of the Human Leucocyte Antigen (HLA) region (8 Mb of chromosome 6p21), in susceptibility to type 1 diabetes has been known since the 1970s [3]. The HLA has multiple roles in T cell selection, antigen presentation and immune responses, all of which can influence the onset and progression of T1D. The HLA region generally can be split into three different parts, class I, class II, and class III (Figs. 1 and 2).

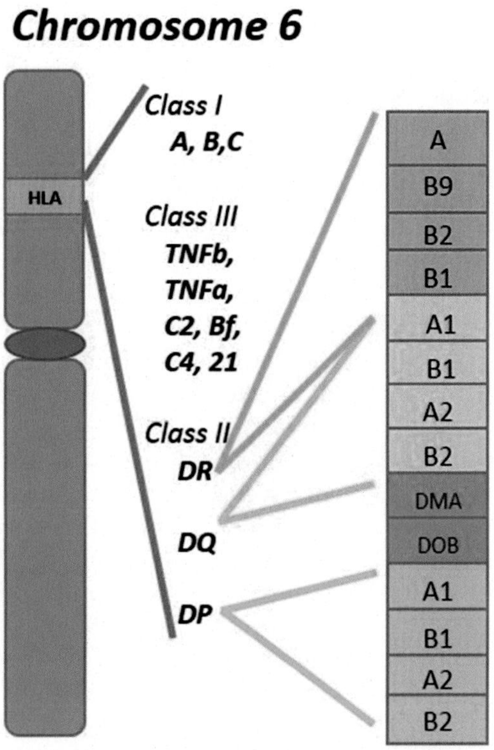

Fig. 1 MHC region of chromosome 6. MHC region is comprised of 3 loci corresponding to HLA Class I, II and III. In HLA Class II region DR, DQ and DP loci are presented. Risk alleles for T1D are located in DR and DQ regions

MHC Class II

Peptide Binding Cleft

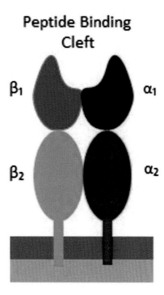

Fig. 2 MHC class II. MHC class II molecules are heterodimeric and consist of two peptides—α and β chains, which include α1, β1 domains encoding binding cleft, and α2, β2—membrane bound domain

The class I region, encoding HLA A, B, and C molecules, is expressed on the cell surface of nucleated cells that are involved in the presentation of endogenous antigens to CD8+ cytotoxic T cells (Tc) and contribute to risk of T1D [11] and the HLA A*24 allele is associated with rapid progression to T1D [12].

The HLA class II region encodes membrane bound proteins expressed on the cell surface of antigen-presenting cells (APCs): B-lymphocytes, macrophages, and dendritic cells that are involved in the processing and presentation of exogenous antigens to CD4+ T helper cells (Th) (Fig. 3) leading to T cell activation. The numerous subsets of T cells are derived from a single T cell precursor and some subsets have the capacity to regulate one another (Fig. 4).

Studies of the pathogenesis of type 1 diabetes proving roles for the HLA as well as effector and regulator T cell population mechanistically have utilized the predominant animal model of type 1 diabetes, the nonobese diabetic (NOD) mouse transgenic for HLA Class II [13].

HLA class II genes contribute to both susceptibility and resistance to T1D; risk is associated with the HLA class II haplotypes *DRB1*04-DQB1*0302* and *DRB1*03-DQB1*02* while the haplotype *DRB1*15-DQB1*0602* is dominantly protective. The risk of

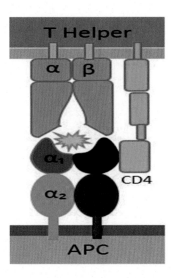

Fig. 3 Antigen presentation. During antigen presentation CD4 receptors of T helper lymphocytes bind to β domain of the HLA Class II molecule that activates the T cell

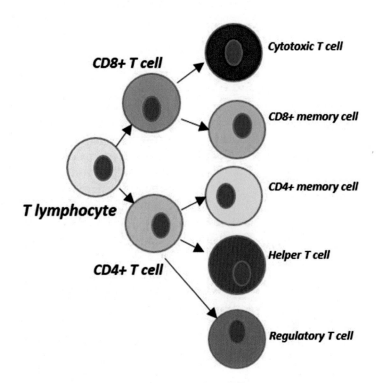

Fig. 4 T cell subsets. CD8+ T cells mature into cytotoxic T cells, CD8+ memory cells; CD4+ T cells mature into helper T cells, regulatory T cells, CD4+ memory cells

Selected T1D associated genes

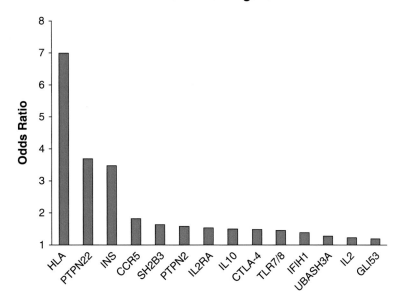

Fig. 5 The relative effects of selected T1D associated genes on susceptibility to T1D (adapted from Todd, 2010) [15]

developing T1D in siblings of affected children varies from 0.3 to 30 % depending on their HLA class II genotype [14].

Additional genetic risk markers were identified in the 1980s and 1990s but the advent of genome-wide association studies (GWAS), conducted since 2007, has allowed identification of approximately 40 additional genes that contribute to susceptibility to T1D (Fig. 5) [15]. Ongoing research now focuses on the biological pathways in immune and beta cells where these T1D associated genes function [16]. Despite these huge advances, all the genes identified to date do not account for the sum of genetic susceptibility. The concept of "missing heritability" has led to a focus on rare variants.

The increasing role of a "diabetogenic" environment was suggested following reports of the rising incidence and decreasing age at diagnosis of T1D while the frequency of the high risk HLA DR3/DR4 genotype is decreasing [17, 18]. A variety of environmental factors such as infections in early life, diet, and early development of the gut microbiome have been implicated in promoting the rising incidence of T1D [19] but none yet unequivocally proven. It is likely that one mechanism by which the environment influences risk of autoimmune diabetes is through epigenetic changes.

3   The Natural History of Type 1 Diabetes

Over the last 20 years, a series of birth cohort studies [20–23] have contributed hugely to our understanding of the natural history of the condition. Islet antibodies are markers of ongoing autoimmune destruction [24] and the best characterized are specific to the islet proteins insulin [25], glutamic acid decarboxylase (GAD) [26], IA-2 [27], and the zinc transporter ZnT8 [28, 29]. The autoimmune process begins very early in life: studies of neonatal diabetes suggest that most cases of diabetes diagnosed before 6 months are unlikely to be autoimmune, but those diagnosed after the age of 6 months have the genetic characteristics of T1D [30]. Antibodies to insulin (generally the first to appear) have been detected as early as 6–12 months of age [31]. Longer term follow-up of birth cohorts in Finland and Germany suggests that there is an explosion of islet autoimmunity in at risk children between the ages of 6 months and 3 years [21, 32]. The techniques to detect islet autoantibodies with high sensitivity and specificity have resulted from decades of collaborative workshops where blinded reference samples are tested in participating laboratories [33] resulting in high quality radioimmunoassays [34] and more recently ELISAs and chemiluminescence assays [35].

Islet autoantibody studies have demonstrated differing rates of progression in individuals positive for multiple islet autoantibodies; many progress rapidly [27] but there is also accumulating evidence for "slow burning" autoimmunity. For instance, within the Bart's Oxford study of type 1 diabetes (http://www.bristol.ac.uk/clinical-sciences/research/diabetes/research/box/), ongoing since 1985, some "at risk" individuals with two or more islet autoantibodies remain diabetes free after 20 years. A relapsing/remitting process of beta cell destruction has been postulated that could help explain differences in rates of progression but as yet there is no widely available marker of beta cell death although an assay to detect demethylated insulin for this purpose has been described [36, 37].

Although the pancreas in type 1 diabetes has been described previously [38, 39], recent histological analysis revealed new insights into the immune cell subsets comprising insulitis with the observation that B cells are present in greater frequency than expected [40]. Further analysis of T1D pancreas has been made possible by the Network for pancreatic organ donors with diabetes (nPOD) initiative (www.jdrfnpod.org). Improving techniques raise the possibility of analysis of single laser captured islet beta cells.

Once diagnosed, the insulin-free "honeymoon period" is variable and there is increasing evidence that some individuals with long-standing diabetes can continue to make low levels of insulin [41]. Large scale testing has been made possible through a straightforward test to detect c-peptide in urine [42].

Future perspectives in type 1 diabetes include improved bio-marker identification to support a number of ongoing clinical trials orchestrated internationally by the TrialNet consortium (www.diabetestrialnet.org).

References

1. Tisch R, Devitt H (1996) Insulin-dependent diabetes mellitus. Cell 85:291–297
2. Butler AE, Galasso R, Meier JJ, Basu R, Rizza RA, Butler PC (2007) Modestly increased beta cell apoptosis but no increased beta cell replication in recent-onset type 1 diabetic patients who died of diabetic ketoacidosis. Diabetologia 50:2323–2331
3. Gillespie K. (2011) The genetics of type 1 diabetes, type 1 diabetes - pathogenesis, genetics and immunotherapy, Prof. David Wagner (Ed.), ISBN: 978-953-307-362-0, Rijeka, Croatia: InTech, Available from: http://www.intechopen.com/books/type-1-diabetes-pathogenesis-genetics-and-immunotherapy/the-geneticsof-type-1-diabetes
4. Karvonen M, Viik-Kajander M, Moltchanova E, Libman I, LaPorte R, Tuomilehto J (2000) Incidence of childhood type 1 diabetes worldwide. Diabetes Mondiale (DiaMond) Project Group. Diabetes Care 23:1516–1526
5. Onkamo P, Vaananen S, Karvonen M, Tuomilehto J (1999) Worldwide increase in incidence of type 1 diabetes - the analysis of the data on published incidence trends. Diabetologia 42:1395–1403
6. Patterson CC, Dahlquist GG, Gyurus E, Green A, Soltesz G (2009) Incidence trends for childhood type 1 diabetes in Europe during 1989-2003 and predicted new cases 2005-20: a multicentre prospective registration study. Lancet 373:2027–2033
7. Redondo MJ, Yu L, Hawa M, Mackenzie T, Pyke DA, Eisenbarth GS et al (2001) Heterogeneity of type I diabetes: analysis of monozygotic twins in Great Britain and the United States. Diabetologia 44:354–362
8. Schenker M, Hummel M, Ferber K, Walter M, Keller E, Albert ED et al (1999) Early expression and high prevalence of islet autoantibodies for DR3/4 heterozygous and DR4/4 homozygous offspring of parents with Type I diabetes: the German BABYDIAB study. Diabetologia 42:671–677
9. Gillespie KM, Gale EA, Bingley PJ (2002) High familial risk and genetic susceptibility in early onset childhood diabetes. Diabetes 51:210–214
10. Gillespie KM, Aitken RJ, Wilson I, Williams AJ, Bingley PJ (2014) Early onset of diabetes in the proband is the major determinant of risk in HLA DR3-DQ2/DR4-DQ8 siblings. Diabetes 63:1041–1047
11. Nejentsev S, Howson JM, Walker NM, Szeszko J, Field SF, Stevens HE et al (2007) Localization of type 1 diabetes susceptibility to the MHC class I genes HLA-B and HLA-A. Nature 450:887–892
12. Mbunwe E, Van der Auwera BJ, Vermeulen I, Demeester S, Van Dalem A, Balti EV et al (2013) HLA-A*24 is an independent predictor of 5-year progression to diabetes in autoantibody-positive first-degree relatives of type 1 diabetic patients. Diabetes 62:1345–1350
13. Mellanby RJ, Phillips JM, Parish NM, Cooke A (2008) Both central and peripheral tolerance mechanisms play roles in diabetes prevention in NOD-E transgenic mice. Autoimmunity 41:383–394
14. Lambert AP, Gillespie KM, Thomson G, Cordell HJ, Todd JA, Gale EAM et al (2004) Absolute risk of childhood-onset type 1 diabetes defined by human leukocyte antigen class II genotype: a population-based study in the United Kingdom. J Clin Endocrinol Metab 89:4037–4043
15. Todd JA (2010) Etiology of type 1 diabetes. Immunity 32:457–467
16. Marroqui L, Santin I, Dos Santos RS, Marselli L, Marchetti P, Eizirik DL (2014) BACH2, a candidate risk gene for type 1 diabetes, regulates apoptosis in pancreatic beta-cells via JNK1 modulation and crosstalk with the candidate gene PTPN2. Diabetes 63:2516–2527
17. Gillespie KM, Bain SC, Barnett AH, Bingley PJ, Christie MR, Gill GV, Gale EA (2004) The rising incidence of childhood type 1 diabetes and reduced contribution of high-risk HLA haplotypes. Lancet 364:1699–1700
18. Fourlanos S, Varney MD, Tait BD, Morahan G, Honeyman MC, Colman PG et al (2008) The rising incidence of type 1 diabetes is accounted for by cases with lower-risk human leukocyte antigen genotypes. Diabetes Care 31:1546–1549

19. Knip M, Simell O (2012) Environmental triggers of type 1 diabetes. Cold Spring Harb Perspect Med 2:a007690

20. Hermann R, Turpeinen H, Laine AP, Veijola R, Knip M, Simell O et al (2003) HLA DR-DQ-encoded genetic determinants of childhood-onset type 1 diabetes in Finland: an analysis of 622 nuclear families. Tissue Antigens 62:162–169

21. Ziegler AG, Rewers M, Simell O, Simell T, Lempainen J, Steck A et al (2013) Seroconversion to multiple islet autoantibodies and risk of progression to diabetes in children. JAMA 309:2473–2479

22. Barker JM, Barriga KJ, Yu L, Miao D, Erlich HA, Norris JM et al (2004) Prediction of Autoantibody positivity and progression to type 1 diabetes: Diabetes Autoimmunity Study in the Young (DAISY). J Clin Endocrinol Metab 89:3896–3902

23. Bingley PJ, Bonifacio E, Williams AJ, Genovese S, Bottazzo GF, Gale EA (1997) Prediction of IDDM in the general population: strategies based on combinations of autoantibody markers. Diabetes 46:1701–1710

24. Bottazzo GF, Dean BM, McNally JM, Mackay EH, Swift PGF, Gamble DR (1985) In situ characterisation of autoimmune phenomena and expression of HLA molecules in the pancreas in diabetic insulitis. N Engl J Med 313:353–360

25. Palmer JP, Asplin CM, Clemons P, Lyen K, Tatpati O, Raghu PK et al (1983) Insulin antibodies in insulin-dependent diabetics before insulin treatment. Science 222:1337–1339

26. Baekkeskov S, Bruining GJ, Molenaar L, Sigurdsson E, Christgau S (1989) Predictive value of Mr 64,000 antibodies for Type 1 (insulin-dependent) diabetes in a childhood population. Diabetologia 32:463A

27. Christie MR, Genovese S, Cassidy D, Bosi E, Brown TJ, Lai M et al (1994) Antibodies to islet 37k antigen but not to glutamate decarboxylase discriminate rapid progression to insulin-dependent diabetes mellitus in endocrine autoimmunity. Diabetologia 43:1254–1259

28. Wenzlau JM, Juhl K, Yu L, Moua O, Sarkar SA, Gottlieb P et al (2007) The cation efflux transporter ZnT8 (Slc30A8) is a major autoantigen in human type 1 diabetes. Proc Natl Acad Sci U S A 104:17040–17045

29. Wenzlau JM, Liu Y, Yu L, Moua O, Fowler KT, Rangasamy S et al (2008) A common nonsynonymous single nucleotide polymorphism in the SLC30A8 gene determines ZnT8 autoantibody specificity in type 1 diabetes. Diabetes 57:2693–2697

30. Edghill EL, Dix RJ, Flanagan SE, Bingley PJ, Hattersley AT, Ellard S et al (2006) HLA genotyping supports a nonautoimmune etiology in patients diagnosed with diabetes under the age of 6 months. Diabetes 55:1895–1898

31. Roll U, Christie MR, Fuchtensbusch M, Payton MA, Hawkes CJ, Ziegler AG (1996) Perinatal autoimmunity in offspring of diabetic parents: the German Multicenter BABY-DIAB Study: detection of humoral immune responses to islet autoantigens in early childhood. Diabetes 145:967–973

32. Parikka V, Nanto-Salonen K, Saarinen M, Simell T, Ilonen J, Hyoty H et al (2012) Early seroconversion and rapidly increasing autoantibody concentrations predict prepubertal manifestation of type 1 diabetes in children at genetic risk. Diabetologia 55:1926–1936

33. Bingley PJ, Williams AJ (2004) Validation of autoantibody assays in type 1 diabetes: workshop programme. Autoimmunity 37:257–260

34. Williams AJ, Bingley PJ, Bonifacio E, Palmer JP, Gale EA (1997) A novel micro-assay for insulin autoantibodies. J Autoimmun 10:473–478

35. Yu L, Dong F, Miao D, Fouts AR, Wenzlau JM, Steck AK (2013) Proinsulin/Insulin autoantibodies measured with electrochemiluminescent assay are the earliest indicator of prediabetic islet autoimmunity. Diabetes Care 36:2266–2270

36. Akirav EM, Lebastchi J, Galvan EM, Henegariu O, Akirav M, Ablamunits V et al (2011) Detection of β cell death in diabetes using differentially methylated circulating DNA. Proc Natl Acad Sci U S A 108:19018–19023

37. Herold KC, Usmani-Brown S, Ghazi T, Lebastchi J, Beam CA, Bellin MD et al (2015) Beta Cell death and dysfunction during type 1 diabetes development in at-risk individuals. J Clin Invest 125:1163–1173

38. Gepts W, DeMey J (1978) Islet cell survival determined by morphology: an immunocytochemical study of the islet of Langerhans in juvenile diabetes mellitus. Diabetes 27:251–261

39. Foulis AK, Liddle CN, Farquharson MA, Richmond JA, Weir RS (1986) The histopathology of the pancreas in Type 1 (insulin-dependent) diabetes mellitus: a 25-year review of deaths in patients under 20 years of age in the United Kingdom. Diabetologia 29:267–274

40. Willcox A, Richardson SJ, Bone AJ, Foulis AK, Morgan NG (2009) Analysis of islet inflammation in human type 1 diabetes. Clin Exp Immunol 155:173–181

41. Oram RA, Jones AG, Besser RE, Knight BA, Shields BM, Brown RJ et al (2014) The majority of patients with long-duration type 1 diabetes are insulin microsecretors and have functioning beta cells. Diabetologia 57:187–191

42. McDonald TJ, Knight BA, Shields BM, Bowman P, Salzmann MB, Hattersley AT (2009) Stability and reproducibility of a single-sample urinary C-peptide/creatinine ratio and its correlation with 24-h urinary C-peptide. Clin Chem 55:2035–2039

Part I

Genes

Methods in Molecular Biology (2016) 1433: 13–20
DOI 10.1007/7651_2015_307
© Springer Science+Business Media New York 2015
Published online: 13 December 2015

Type 1 Diabetes High-Risk HLA Class II Determination by Polymerase Chain Reaction Sequence-Specific Primers

Rachel J. Aitken, Georgina L. Mortimer, and Kathleen M. Gillespie

Abstract

The only strategy to select individuals at increased risk for type 1 diabetes for primary prevention trials is through genetic risk assessment. While genome-wide association studies have identified more than 40 loci associated with type 1 diabetes, the single most important genetic determinants lie within the human leucocyte antigen gene family on chromosome 6.

In this chapter we describe a protocol for a straightforward, cheap strategy to determine HLA class II mediated risk of type 1 diabetes. This method has proved robust for genotyping whole-genome-amplified DNA as well as DNA extracted directly from human tissues.

Keywords: HLA class II, HLA class 1, Genetic risk

1 Introduction

HLA class II presents self-peptides to the immune system and autoimmunity results from impaired tolerance to self-antigens. A fundamental role for HLA class II in susceptibility to type 1 diabetes (T1D) is therefore not surprising. An association between HLA and T1D was originally demonstrated in the 1970s [1, 2] with several major steps forward in the ensuing decades when the importance of HLA class II alleles was described [3] the relationship between age at onset of diabetes and HLA class II-mediated risk [4–6], and the hierarchy of HLA class II-mediated risk in type 1 diabetes [7].

The method described here has been adapted from the original "Phototyping" method for discrimination of HLA genotype by Bunce and colleagues [8]. It allows identification of all alleles of HLA *DRB1* facilitating identification of homozygosity and heterozygosity as well as selected alleles from HLA *DQA1* and *DQB1*. Overall the highest risk diplotype is HLA *DRB1*03-DQB1*0201/DRB1*04-DQB1*0302.*

2   Materials

2.1   DNA

DNA extracted from human tissues: This includes tissues extracted from fixed tissues using appropriate extraction kits and whole-genome-amplified DNA (**Note 1**).

2.2   Reagents and Supplies

PCR reaction

1. Oligonucleotide primers specific for HLA *DRB1*, *DQA1*, and *DQB1* as described in Tables 1 and 2 (Fig. 1). Make up to a standardized concentration of 100 pmol/μl and store in aliquots at −20 °C. A control primer set (any robust set will suffice) must be added to each allele specific primer set.

2. Taq polymerase and PCR Buffer, MgCl$_2$ (usually supplied together). Go Taq and buffer from Promega are used in the method described (**Note 2**).

3. dNTPs commercially available

4. 96-Well PCR plates

5. 96-Well thermocycler

6. 10× TBE buffer: 108 g Tris and 55 g boric acid in 800 ml dH$_2$O. Add 40 ml 0.5 M Na$_2$EDTA (pH 8.0). Adjust volume to 1 l. Store at room temperature.

7. Agarose gel electrophoresis tank, gel-forming tray, and powerpack

Agarose Gel Electrophoresis

1. 2 % agarose gel: 4 g Ultrapure Agarose, 200 ml 1× TBE buffer

2. Ethidium bromide/Midori green or alternative

Table 1
HLA DRB1 primer sequences. DR2 has been split to DR15 and 16, DR5 to DR11 and 12, and DR6 to DR13 and 14

Mix No.	Gene	Sequence (5′–3′)
1	DR1F	TTG TGG CAG CTT AAG TTT GAA T
	DR1R1	CTG CAC TGT GAA GCT CTC AC
	DR1R2	CTG CAC TGT GAA GCT CTC CA
2	DR15F	TCC TGT GGC AGC CTA AGA G
	DR15R	CCG CGC CTG CTC CAG GAT
3	DR16F	TCC TGT GGC AGC CTA AGA G
	DR16R	AGG TGT CCA CCG CGG CG
4	DR3F	GTT TCT TGG AGT ACT CTA CGT C
	DR3R	TGC AGT AGT TGT CCA CCC G

(continued)

Table 1
(continued)

Mix No.	Gene	Sequence (5′–3′)
5	DR4F	GTT TCT TGG AGC AGG TTA AAC A
	DR4R1	CTG CAC TGT GAA GCT CTC AC
	DR4R2	CTG CAC TGT GAA GCT CTC CA
6	DR11F	GTT TCT TGG AGT ACT CTA CGT C
	DR11R2	CTG GCT GTT CCA GTA CTC CT
7	DR12F	AGT ACT CTA CGG GTG AGT GTT
	DR12R	CAC TGT GAA GCT CTC CAC AG
8	DR13F1	TAC TTC CAT AAC CAG GAG GAG A
	DR13F2	GTT TCT TGG AGT ACT CTA CGT C
	DR13R1	CCC GCT CGT CTT CCA GGA T
	DR13R2	TGT TCC AGT ACT CGG CGC T
	DR13R3	CCC GCC TGT CTT CCA GGA A
9	DR14F1	GTT TCT TGG AGT ACT CTA CGT C
	DR14F2	AGT ACT CTA CGG GTG AGT GTT
	DR14R	TCT GCA ATA GGT GTC CAC CT
10	DR7F	CCT GTG GCA GGG TAA GTA TA
	DR7R	CCC GTA GTT GTG TCT GCA CAC
11	DR8F	AGT ACT CTA CGG GTG AGT GTT
	DR8R	TGT TCC AGT ACT CGG CGC T
	DR8R	GCT GTT CCA GTA CTC GGC AT
12	DR9F	GTT TCT TGA AGC AGG ATA AGT TT
	DR9R	CCC GTA GTT GTG TCT GCA CAC
13	DR10F	CGG TTG CTG GAA AGA CGC G
	DR10R	CTG CAC TGT GAA GCT CTC AC

Table 2
HLA DQB1 primer sequences

Mix No.	Gene	Sequence (5′–3′)
1	DQB1*02 F	GTGCGTCTTGTGAGCAGAAG
	DQB1*02 R	GTAGTTGTGTCTGCACACCC
2	DQB1*0201 F	GTCCGGTGGTTTCGGAATGA
	DQB1*0201 R	TGCTCTGGGCAGATTCAGAT
3	DQB1*0301/0304 F	GACGGAGCGCGTGCGTTA
	DQB1*0301/0304 R	CGTGCGGAGCTCCAACTG
4	DQB1*0304 F	TTTCGTGCTCCAGTTTAAGGC
	DQB1*0304 R	TGGCTGTTCCAGTACTCGGCGG

(continued)

Table 2
(continued)

Mix No.	Gene	Sequence (5′–3′)
5	DQB1*0302 F DQB1*0302 R	GTGCGTCTTGTGACCAGATA CTGTTCCAGTACTCGGCGG
6	DQB1*0307 F DQB1*0307 R	CCCGCAGAGGATTTCGTGTA CCCCAGCGGCGTCACCA
7	DQB1*0303 F DQB1*0303 R	GACCGAGCGCGTGCGTCT CTGTTCCAGTACTCGGCGT
8	DQB1*0305 F DQB1*0305 R	GCTACTTCACCAACGGGACC TGCACACCGTGTCCAACTC
9	DQB1*0401/0402 F DQB1*0401/0402 R	CTACTTCACCAACGGGACC TGGTAGTTGTGTCTGCATACG
10	DQB1*0501 DQB1*0501	ACGGAGCGCGTGCGGGG GCTGTTCCAGTACTCGGCAA
11	DQB1*0502 DQB1*0502	TGCGGGGTGTGACCAGAC TGTTCCAGTACTCGGCGCT
12	DQB1*0503 DQB1*0503	TGCGGGGTGTGACCAGAC GCGGCGTCACCGCCCGA
13	DQB1*0601 DQB1*0601	TTTCGTGCTCCAGTTTAAGGC CCGCGGAACGCCAGCTC
14	DQB1*0602/10/13 DQB1*0602/10/13	CCCGCAGAGGATTTCGTGTT CCTGCGGCGTCACCGCG
15	DQB1*0603/7 DQB1*0603/7	GGAGCGCGTGCGTCTTGTA GCTGTTCCAGTACTCGGCAT
16	DQB1*0603/8/12 DQB1*0603/8/12	GGAGCGCGTGCGTCTTGTA AACTCCGCCCGGGTCCC
17	DQB1*0603-05 0607-09 0612 DQB1*0603-05 0607-09 0612	GGAGCGCGTGCGTCTTGTA TGCACACCGTGTCCAACTC

Example of HLA DQB1 genotyping

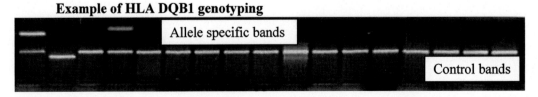

Fig. 1 A typical example of an HLA DQB1 genotyping result. In this case the sample is positive for *HLA DQB1*02 (*0201*) and *DQB1*0302. Note that in lane 2 the target *HLA DQB1*0201* allele is the larger band observed. The control band set used in this experiment is Human growth hormone; F5′CCAGCTCAAGGATCCCAA and R5′ CACCCATTACCCAAGAGCTTA

3   Methods

3.1   Set Up PCR Reaction

1. Label a 96-well plate for the number of reactions to be carried out; for instance a full HLA *DRB1* type requires 14 reactions per sample.

2. Following the worksheet provided in Table 3; briefly add the DNA (**Note 1**) to the appropriate well. Ensure that positive and negative (dH$_2$O) controls are included.

3. Make up a "cocktail" of all other reagents for the required number of reactions

Table 3
Work sheet for typing HLA class II DRB1 and DQB1

PCR mix	DNA (20 ng/µl)/whole-genome-amplified DNA		
		Per *x* reactions	
	Per reaction (10 µl)	DR (×16 to include controls)	DQB (×20 to include controls)
GoTaq G2 Green buffer	2	32	40
dNTP's 1 mM	0.3	4.8	6
25 mM MgCl$_2$	0.6	9.6	12
DNA	0.6	9.6	12
dH$_2$O	1.45	23.2	29
GoTaq G2 5 U/µl HLA	0.05	0.8	1
Primer (F&R) + control (FdR)	5		
Samples to type	*Comments*	*Samples to type*	*Comments*
1		13	
2		14	
3		15	
4		16	
5		17	
6		18	
7		19	
8		20	
9		21	
10		22	
11		23	
12		24	

4. Vortex to mix

5. Add the appropriate volume of "cocktail" to each well.

6. Seal plate

7. Spin down in an appropriate centrifuge or gently tap the plate to ensure that all reagents are mixed.

3.2 Thermal Cycler Program

Select the following touchdown PCR program on the thermal cycler

Steps	Temperature (°C)	Time (sec.)	Action
1 cycle	96	60	Denaturation
5 cycles	96	25	Denaturation
	70	50	Annealing
	72	45	Extension
21 cycles	96	25	Denaturation
	65	50	Annealing
	72	45	Extension
4 cycles	96	25	Denaturation
	55	60	Annealing
	72	120	Extension
Hold	4	Specify time	

3.3 Agarose Gel Electrophoresis

1. Weigh out 4 g of agarose into a conical flask. Add 200 mL of 1× TBE, and swirl to mix (**Notes 3** and **4**).

2. Microwave for about 2 min to completely dissolve the agarose. Stop after 45 s to give agarose mix a swirl taking precautions not to allow agarose to boil over causing a burn.

3. Ensure the agarose is completely dissolved. If not replace in the microwave until dissolved monitoring carefully.

4. Cool the agarose by swirling the conical flask under a running cold tap until the gel is about 60 °C.

5. Add 2 μL of ethidium bromide (10 mg/mL) or alternative DNA stain for instance Midori green and swirl to mix (**Note 5**).

6. Pour the gel into a pre-prepared gel tank with appropriate sized gel combs on a level surface.

7. Allow to set for 1 h.

8. Pour 1× TBE buffer into the gel tank to submerge the gel (**Note 6**).

9. Remove combs ensuring that wells are well formed

10. Pipette 10 μl of PCR product into the gel and run at 100–115 V for approximately an hour (**Note 7**).

11. Visualise the gel on an appropriate gel documentation system.

4   Notes

1. This method works well on DNA from tissues or whole-genome-amplified DNA but ensure that whole-genome-amplified DNA is appropriately diluted.

2. Go Taq (Promega) has been used in this protocol but other enzymes should also work. This enzyme comes with a coloured buffer with sufficient density that a gel loading dye is not required. An alternative to this addition of sucrose creosol red which acts as a density agent for gel electrophoresis but can be added to the PCR reaction without any negative affect. This saves time adding a loading buffer for electrophoresis. The recipe is as follows: 100 mM Creosol Red dye: dissolve 0.4 g creosol red in 10 ml ddH$_2$O, vortex to mix. Before setting up the PCR reaction a loading dye can be generated as follows: 60 % Sucrose/1 mM Creosol Red (50 ml): Dissolve 30 g sucrose in 50 ml autoclaved deionized H$_2$O, add 500 µl of 100 mM creosol red, vortex to mix. Aliquot 1 ml into 1.5 ml Eppendorfs, and store at −20 °C. Add 1.5 µl during the setup of a 10 µl reaction

3. Use a large container, as long as it fits in the microwave, because the agarose boils over easily.

4. The volume of gel can be scaled up depending on the number of samples to be analyzed and the gel equipment available.

5. Ethidium bromide is mutagenic and should be handled with caution. Contaminated tips can be disposed of in a dedicated ethidium bromide waste container. Alternative DNA stains are increasingly used.

6. The gel must be run in the same buffer as used to make up the gel.

7. DNA is negatively charged and will run towards the anode.

Acknowledgement

This work was supported by Diabetes UK grants to KMG.

References

1. Nerup J, Platz P, Andersen OO et al (1974) HL-A antigens and diabetes mellitus. Lancet 2:864–866

2. Cudworth AG, Woodrow JC (1975) Evidence for HL-A-linked genes in 'juvenile' diabetes mellitus. Br Med J 3:133–135

3. Todd JA, Bell JI, McDevitt HO (1987) HLA-DQ beta gene contributes to susceptibility and resistance to insulin-dependent diabetes mellitus. Nature 329:599–604

4. Caillat-Zucman S, Garchon HJ, Timsit J, Assan R, Boitard C, Djilali-Saiah I, Bougnères P, Bach

JF (1992) Age-dependent HLA genetic heterogeneity of type 1 insulin-dependent diabetes mellitus. J Clin Invest 90:2242–2250

5. Gillespie KM, Gale EA, Bingley PJ (2002) High familial risk and genetic susceptibility in early-onset childhood diabetes. Diabetes 51:210–214

6. Gillespie KM, Aitken RJ, Wilson I et al (2014) Early onset of diabetes in the proband is the major determinant of risk in HLA DR3-DQ2/DR4-DQ8 siblings. Diabetes 63:1041–1047

7. Lambert AP, Gillespie KM, Thomson G et al (2004) Absolute risk of childhood-onset type 1 diabetes defined by human leukocyte antigen class II genotype: a population-based study in the United Kingdom. J Clin Endocrinol Metab 89:4037–4043

8. Bunce M, O'Neill CM, Barnardo MC et al (1995) Phototyping: comprehensive DNA typing for HLA-A, B, C, DRB1, DRB3, DRB4, DRB5 & DQB1 by PCR with 144 primer mixes utilizing sequence-specific primers (PCR-SSP). Tissue Antigens 46:355–367

Methods in Molecular Biology (2016) 1433: 21–54
DOI 10.1007/7651_2015_291
© Springer Science+Business Media New York 2015
Published online: 03 March 2016

Pancreatic Beta Cell Survival and Signaling Pathways: Effects of Type 1 Diabetes-Associated Genetic Variants

Izortze Santin, Reinaldo S. Dos Santos, and Decio L. Eizirik

Abstract

Type 1 diabetes (T1D) is a complex autoimmune disease in which pancreatic beta cells are specifically destroyed by the immune system. The disease has an important genetic component and more than 50 *loci* across the genome have been associated with risk of developing T1D. The molecular mechanisms by which these putative T1D candidate genes modulate disease risk, however, remain poorly characterized and little is known about their effects in pancreatic beta cells. Functional studies in in vitro models of pancreatic beta cells, based on techniques to inhibit or overexpress T1D candidate genes, allow the functional characterization of several T1D candidate genes. This requires a multistage procedure comprising two major steps, namely accurate selection of genes of potential interest and then in vitro and/or in vivo mechanistic approaches to characterize their role in pancreatic beta cell dysfunction and death in T1D. This chapter details the methods and settings used by our groups to characterize the role of T1D candidate genes on pancreatic beta cell survival and signaling pathways, with particular focus on potentially relevant pathways in the pathogenesis of T1D, i.e., inflammation and innate immune responses, apoptosis, beta cell metabolism and function.

Keywords: Type 1 diabetes, Diabetes candidate genes, Pancreatic beta cells, Pancreatic islets, Inflammation, Apoptosis, Cytokines, Small interfering RNA (siRNA), Overexpression vector

1 Introduction

Genome wide association studies (GWAS) have identified more than 50 genomic regions associated with type 1 diabetes (T1D) risk [1]. Many candidate genes have been proposed within these regions but few have been confirmed as true etiological genes, and the molecular mechanisms by which these candidate genes modulate disease risk remain poorly characterized [1]. The big challenge in the field is to define how candidate genes for T1D interact with the environment and modulate the development of diabetes.

Chronic and excessive inflammation contributes to tissue damage in diseases such as lupus erythematosus, rheumatoid arthritis and T1D. There is thus a major unmet need for therapies that modulate excessive inflammation without affecting immune defenses against invading pathogens. Achieving this goal will require a comprehensive understanding of molecular pathways and mechanisms that regulate the expression of pro-inflammatory genes [2], and the role of the individual's genetic background in these processes. An increased innate inflammatory state, due to inheritance of potentiating genetic variants in immune pathways that is independent of auto-antibodies, HLA status and disease progression, is likely to exist in T1D families [3]. Until recently nearly all studies dealing with genetic risk for T1D focused on the putative role of candidate genes on the immune system [4]. We and others have shown, however, that T1D genes regulate pathways that may be crucial for the pathogenesis of diabetes at the pancreatic beta cell level [5–12]. These observations suggest that four major genetically regulated pathways play a role in pancreatic beta cell dysfunction and death in T1D, e.g., innate immunity and antiviral activity [6, 10], and pathways related to beta cell phenotype and susceptibility to pro-apoptotic stimuli [7, 8, 10–12].

Assessment of the real contribution of a given T1D susceptibility gene to the pathogenesis of the disease requires a multistage procedure, combining the use of genetic association data, gene expression profiles and other transcriptomic data, advanced bioinformatics techniques, and functional assays in in vitro and in vivo models [8, 10, 13].

1.1 Selection of Genes for Functional Studies

Genetic risk to develop T1D has been associated with several regions along the genome [1, 4]. Causal candidate genes have been proposed based on the localization of the association signal (positional candidates) or on their function (functional candidates), but most of these regions contain more than one protein-coding gene and several noncoding features, such as long noncoding RNAs (lncRNAs) or microRNAs (miRNAs). Thus, selecting candidate genes for functional studies is not trivial and requires an accurate multistage selection procedure.

In our laboratory we are following a "minimalist" selection procedure composed of three steps that will be discussed in detail here: (a) Identification of genes falling into GWAS association signals; (b) Determination on whether these genes are expressed in human and rodent beta cells; (c) Identification of genes with a potential relevant function in three key signaling pathways for T1D pathogenesis: inflammation/innate immunity, apoptosis and beta cell metabolism/function (Fig. 1). The reader is referred to additional and elegant approaches for candidate gene selection based on in silico "phenome-interactome analysis" and integration of GWAS data with protein–protein interactions to construct biological networks of putative relevance for T1D [8–10, 13]. Moreover, the relevance of understanding candidate genes inside functional pathways has led to the development of novel tools that include pipelines to test for enrichment of T1D association signals among the targets of a given set of transcription factors [14]. To select potential causal candidate genes, other approaches take advantage of the great amount of available GWAS data to identify functionally

MULTI-STAGE SELECTION PROCEDURE
Selection of T1D candidate genes for functional characterization

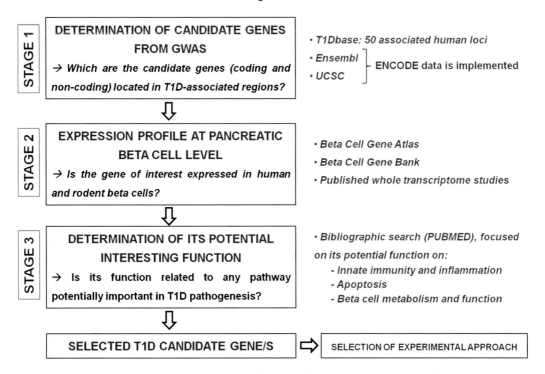

Fig. 1 Multistage selection procedure to identify T1D candidate genes for functional characterization in pancreatic beta cells. The selection of T1D candidate genes for functional studies in pancreatic beta comprises three main steps: (1) Identification of candidate genes located in T1D-associated genomic regions; (2) Evaluation of their expression profiles in rodent and human pancreatic beta cells; (3) Mechanistic determination of their potential role in T1D pathogenesis

related genes spanning multiple GWAS *loci* by using genome-scale shared-function networks [15]. By using a similar strategy, in which a gene pathway-based analysis method is applied to summary statistics of GWAS, novel genetic associations and potential novel causal genes have been identified [16].

1.1.1 Identification of Genes Falling into GWAS Association Signals

The T1Dbase (http://www.t1dbase.org/page/Welcome/display) is a web-based resource focused on the genetics and genomics of type 1 diabetes and containing a table of human, mouse, and rat *loci* associated with T1D [17]. Associated regions are defined by starting at the associated variant of a given GWAS publication and extending out the region by ± 0.1 cM. The newly extended region is further investigated for variants of genome-wide significance; these steps are repeated until no more variants reach a significance of 1E−06 within the cited study. Thus, this table shows all genes falling into association peaks based on the data published by several GWAS [18–21]. The T1Dbase is a valuable tool to search for potential functional candidate genes in T1D-associated genomic regions, but it is important to keep in mind that the protein-coding and noncoding genes/regions were extracted from Ensembl release 73 (we are currently at release 75) and that the latest update of the database was done in 2011 [22]. Thus, and in order to have an updated information, it is important to corroborate the data obtained from the T1Dbase using other databases such as Ensembl (http://www.ensembl.org/index.html) or UCSC Genome Browser (http://genome.ucsc.edu/) in which recent genomic information obtained by the ENCODE project has been implemented.

1.1.2 Expression Profile in Pancreatic Beta Cells

There are several online resources currently available to check whether a candidate gene is expressed in pancreatic beta cells. Among them, the EuroDia database (http://eurodia.vital-it.ch) contains a collection of global gene expression determinations performed on beta-cells of three organisms (human, mouse, and rat) [23]. The Gene Expression Data Analysis Interface (GEDAI) has been developed to support this database in which, in addition to the expression data repository, several tools for mining the data are available. At the time of publication (2010), the EuroDia database contained 38 curated experiments (441 hybridizations), thirteen of which were produced by members of the EC-supported EuroDia project. The database has been periodically updated and presently expression data of 50 experiments comprising 684 hybridizations are available.

The Beta Cell Gene Atlas (http://www.t1dbase.org/page/AtlasHome) is a web-accessible database in which basal expression of genes from different beta cell-related sources, including human, mouse, and rat pancreatic beta cells, islets and whole pancreas, as well as clonal beta cell lines, can be consulted [24]. The Beta Cell Gene Atlas is a collection of public microarray data generated from 131 array analyses derived from 28 experiments published between 2001 and 2006. This has not been updated in recent years, and other resources should be checked to make it sure that the gene is indeed expressed in beta cells. Another open access source is the Beta Cell Gene Bank (http://betacellgenebank.ulb.ac.be/cgi-bin/dispatcher.cgi/BCGB_Enter/display) in which array expression profiles from rodent and human pancreatic beta cells are found, both at basal condition and after exposure to a given insult (e.g., exposure to pro-inflammatory cytokines, transfection with synthetic viral double stranded RNA (Polyinosinic–polycytidylic acid; PIC) or infection with potentially diabetogenic viruses); information (based on manual curation) is provided for >500 genes. The expression data available in this database is based on microarray experiments published between 2003 and 2005 [25–28], and thus updated resources should be checked in order to confirm the expression data of a given gene. The fate of the databases described above, i.e., cessation of updating after some years of development is unfortunately common in the field, reflecting the lack of interest of financing agencies in supporting these crucial resources on a long-term basis.

The Human Islet Regulome Browser (http://www. isletregulome.org/) is a recently developed tool in which transcripts of human islets, chromatin states, and transcription factor binding sites have been mapped [29]. This very elegant resource enables data downloads and online visualization at desired levels of resolution for islet transcription factor binding sites, chromatin states, motifs, enhancer clusters, and genome-wide significant *p*-values for association with type 2 diabetes and fasting glycemia.

In recent years the advent of next-generation RNA sequencing (RNAseq) has provided an unbiased and high-throughput method for determining the whole transcriptome, allowing the identification of novel transcripts in several cell types and tissues [30–33], including the original RNAseq study of human pancreatic islets [34]. Recent beta cell transcriptome studies based on RNAseq used enriched beta cell preparations (both rodent and human) and include studies establishing the gene expression profile for each islet cell subtype [30, 35, 36] and studies assessing the effect of immune (e.g., cytokine exposure) or metabolic (e.g., palmitate exposure) stress on gene expression patterns in whole human islets [34, 37]. The expression data obtained by these RNA sequencing analyses can be scrutinized to check whether the gene of interest is expressed in pancreatic islets or in beta cells, basally or following exposure to inflammation- or metabolic-mediated stress. Publication of most of these studies were accompanied by depositing the raw data in the Gene Expression Omnibus (http://www.ncbi.nlm. nih.gov/geo/) database, an open access resource in which gene expression profiles of array- and sequencing-based experiments can be consulted.

In order to validate the expression data publically available, expression of the gene of interest should be confirmed in both rodent and human pancreatic beta cell samples using real-time PCR, Western blot and histology. In these steps the human islet data is the "golden standard," i.e., if a gene of interest is present in human but not rodent beta cells we go on with the study, but not in the other way around (for instance, caspase-12 was shown to be a key regulator of ER stress-induced apoptosis in rodent cells, including beta cells, but it is not expressed in most human beings). Thus, interspecies variation reinforces the need for human models. Once the expression of a candidate gene is confirmed in human beta cells, the next step involves exposure of human islets or the recently developed human cell line EndoC-βH1 [38] to siRNAs or viral vectors to respectively downregulate or upregulate expression of the candidate gene, and then treatment with pro-inflammatory cytokines, intracellular double stranded RNA (dsRNA is a by-product generated during replication and transcription of both RNA and DNA viruses, and an efficient inducer of apoptosis, type I interferons, and other cytokines/chemokines important for the host immune response to viral infection) or actual viral infection [6, 7, 11, 12, 39], as described below. We aim to model in vitro and under well-controlled conditions the putative genetic/environmental interactions that may take place in early T1D, with the limitation of addressing only one cell type and not the whole organism.

1.1.3 Identification of Genes with a Potential Relevant Function for T1D Pathogenesis

In order to select candidate genes with a potential effect on inflammation/innate immunity, apoptosis and/or beta cell metabolism and function [5–12], a systematic bibliographic search using appropriate "search terms" is performed in PUBMED (http://www.ncbi.nlm.nih.gov/pubmed). The goal is to answer three key questions: (1) What is the function of the candidate gene of interest? (2) Are there publications analyzing its role in pancreatic beta cells? (3) Are there publications linking the gene of interest with innate immunity/inflammation, apoptosis or cell metabolism and function in beta cells or other cell types?

1.1.4 Limitations of the Selection Procedure and Selection of the Experimental Approach

As mentioned above, the genomic regions associated with T1D as defined by GWAS are usually very large and comprise more than one protein-coding gene. Moreover, protein-coding genes in the T1D-associated regions often coexist with other noncoding genes, such as miRNAs and lncRNAs, that may have a role in disease pathogenesis [40–42]. Indeed, a study overlapping islet-expressed miRNAs against T2D association data identified several T2D association signals in target genes of islet-expressed miRNAs [41]. In addition, a recent work has described a set of differentially expressed miRNAs that stratify T1D patients and nondiabetic individuals [42]. Interestingly, the differentially expressed miRNAs are predicted to regulate the expression of several T1D candidate genes that are modulated by pro-inflammatory cytokines in human pancreatic islets [42].

In short, selection of the "right candidate gene" for functional studies is a challenging procedure that risk leading to the selection of a "wrong gene" without a relevant role in the disease. As observed in the T1Dbase (Fig. 2), and independently of the number of genes located in the associated region, one or two causal genes are usually proposed based on their function and/or on the genomic position of the associated polymorphism. In some cases the proposed causal gene is the real etiological gene in the region, with a well-defined functional effect in the pathogenesis of T1D. Good examples of this are the HLA-DRβ1 gene in region 11p15.5 and the IFIH1 gene in region 2q24.2 [6, 43, 44]. In many cases, however, the downstream functional effects of the proposed candidate gene remain to be confirmed. Thus, the other coding and

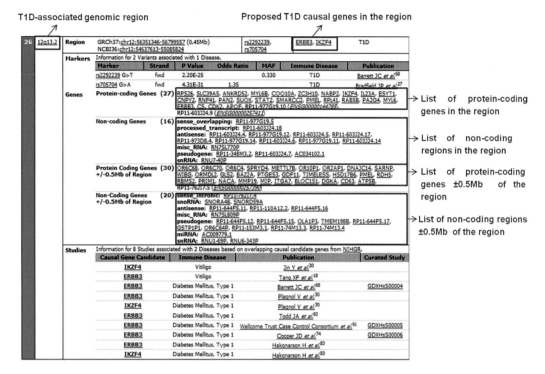

Fig. 2 Interface of a given T1D-associated genomic region from the table of human T1D *loci* in the Beta Cell Gene Bank database. A table with the human T1D *loci* is available at the Beta Cell Gene Bank database (http://betacellgenebank.ulb.ac.be/cgi-bin/dispatcher.cgi/BCGB_Enter/display). As shown in the example, different information are available for each T1D-associated region: proposed T1D candidate genes (based on both position and function), list of other protein-coding and noncoding genes/regions falling under the association peak, and a list of protein-coding and noncoding genes/regions located at a maximum of ± 0.5 Mb from the association signal

noncoding genes in the region should not be excluded as potential causal genes.

It must be also taken into account that several T1D-associated variants do not necessarily affect the expression levels of nearby genes (*cis* eQTLs), and that the associated SNP may have an effect in one or more genes located several megabases away from the association signal (*trans* eQTLs) [29, 45]. A sobering example of this is the recent description that the obesity-associated noncoding sequences within FTO, a candidate gene for obesity and T2D [46, 47], are functionally connected with the homeobox gene IRX3, located at a megabase distance [48]. Indeed, functional studies suggest that the obesity-associated interval is related to the regulatory landscape of IRX3 [48], and not to FTO itself, as previously believed. Expression quantitative trait locus (eQTL; genomic regions that regulate RNA expression level) mapping offers a powerful approach to elucidate the genetic components regulating gene expression [45, 49–51]. Most eQTLs studies have been performed with RNA isolated from blood cells due to the difficulty in obtaining human samples from tissues of difficult access (e.g., pancreatic islets). eQTLs are, however, highly dependent on the specific cell type studied, and eQTL data obtained in short-lived blood circulating cells may not provide adequate information on the impact of the polymorphism in the very long-lived pancreatic beta cells [52]. Moreover, eQTLs are also dependent on the in vitro conditions to which a cell is exposed: e.g., cell exposure to immune activators such as viruses, cytokines, and bacterial products induce major changes in eQTL [53, 54] and these activators should be selected according to the pathogenic mechanism under study. For instance, when studying the impact of the genetic background of an individual in the induction of toxic shock, it may be relevant to determine both basal and lipototoxin-modulated eQTLs in immune cells. Thus, and in the case of diabetes, definition of genetic variants that regulate expression of potential relevant genes for diabetes must rely on eQTL evaluated in human islets and immune cells under basal condition and after exposure to relevant stimuli such as pro-inflammatory cytokines or dsRNA for T1D or free fatty acids or hyperglycemia for T2D (in the case of T2D, human islets and target tissues of insulin, such as liver, muscle and fat, should be evaluated). To reach this aim, a sample size of around 750 preparations is required to enable a statistical power of 80 % to detect *cis* eQTLs with a low minor allele frequency (MAF) [52]; only a global collaborative effort from different human islet-isolating groups around the world can allow the collection and treatment of these large number of human islet preparations under well-standardized conditions.

Another potential issue when performing functional studies of candidate genes for T1D is the risk of selecting the wrong tissue (e.g., pancreatic beta cells or the immune system). T1D is a disease in which the immune system, pancreatic beta cells, and most probably the gut and the microbiome interact [55–57], and it is conceivable that some candidate genes exert composite or even antagonistic effects on these different systems. Deletion/overexpression studies in mouse models may address this issue, but there is an increasing concern regarding the extrapolation of basic mechanisms of inflammation/immunity from mice to humans [58].

Finally, it is important to keep in mind that triggering of complex diseases such as T1D are the consequence of the sum of variations in many genes in cross-talk with multiple environmental factors. Thus, and in order to decipher how a specific genetic background contributes to disease pathogenesis, we need to understand the combined effects of associated genes and their products in the context of interacting functional pathways [10, 59]. For Mendelian disorders in which a mutation in a single gene is sufficient to generate a given pathogenic phenotype, study of a gene-at-a-time is a suitable approach; in complex diseases, however, in which a gene's effect is usually pleiotropic, context-dependent and contingent on other genes from the same functional pathway, pathway analyses should be performed in order to infer key regulating genes in potential pathogenic pathways [10, 59, 60].

After performing the multistage selection procedure described in Fig. 1, a T1D candidate gene will be chosen for further functional studies in pancreatic beta cell models. As mentioned above, T1D candidate genes may not act as isolated elements in disease pathogenesis and when analyzing the functional effect of a given T1D gene we should consider its potential interaction with other T1D candidate genes [10]. To this aim, there are several online resources that can be used to predict possible interactions between a given T1D candidate gene and other (candidate) genes [61]. One of them is GeneMANIA (http://www.genemania.org), a free online user-friendly web interface for generating hypotheses about gene function, prioritizing genes for functional assays and finding potential interactions between genes based on co-expression, physical interaction, and co-localization data, among others [62] (Fig. 3). Through this online prediction tool we can identify potential interactions between (candidate) genes. This allows the design of experimental procedures to analyze the effect of one gene in combination with others at the pancreatic beta cell level (e.g., by inhibition or overexpression of two or more candidate genes simultaneously). In addition, using this kind of prediction tools we can define potential pathway regulators that allow the analysis of a whole pathogenic pathway via modulation of a single regulatory gene. Indeed, recent works in which analysis of gene networks and protein–protein interaction data were combined with GWAS association data have identified disease-relevant biological gene-networks that are enriched with T1D candidate genes or are directly controlled by one of them [9, 60]. For example, a recent work revealed that a T1D-associated gene network implicated in antiviral responses is partially regulated by the T1D candidate gene EBI2 [60].

To choose an appropriate experimental approach for functional studies (i.e., overexpression or inhibition), it is important to check the predicted functional effect of the associated polymorphism in the expression and function of the gene of interest. To this aim, there are several user-friendly online prediction tools that can be used, such as FastSNP (http://fastsnp.ibms.sinica.edu.tw) and SNPinfo (http://snpinfo.niehs.nih.gov/snpinfo/snpfunc.htm), but unfortunately this crucial information is often not available.

To assess more closely the putative effect of a disease-associated variants using experimental models (e.g., T1D-associated polymorphisms in pancreatic beta cell models), a new technique that allows genome editing in eukaryotic cells has been developed [63]. This technique named CRISPR-Cas9 provides an effective and simple method for making small edits in the genome, such as the

a

b

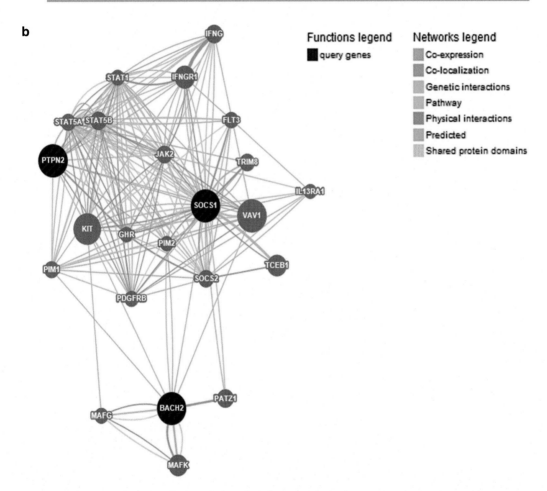

Fig. 3 The online free bioinformatic tool GeneMania allows the finding of potential interactions between genes. Based on data from co-expression, co-localization, genetic interactions, pathway analysis, physical interaction, in silico prediction, and shared protein domains, GeneMania (http://www.genemania.org) generates potential interaction nets. (**a**) The interface of GeneMania allows to query several genes selected from different species. (**b**) After analysis, GeneMania provides a schematic interpretation of potential interactions between queried genes and others. As shown in the legend, query genes are represented by *black circles*, other genes in the network in *grey*, while circles and interactions between genes are defined using *colored lines*. Interestingly, the suggested interaction between the T1D candidate genes BACH2 and PTPN2 has been recently experimentally confirmed [12]

introduction of single nucleotide polymorphisms (SNPs) for probing causal genetic variations in cell lines [63]. CRISPR-Cas9 is a microbial adaptive immune system that uses RNA-guided nucleases to cleave foreign genetic elements. Cas9 promotes genome editing by stimulating a DNA double-stranded break (DSB) at a target genomic locus that, upon cleavage of Cas9, undergoes one of the major pathways for DNA damage repair that can be used to achieve a desired editing outcome. This technique is divided in four steps: (a) In silico design; (b) Reagent construction; (c) Functional validation; (d) Clonal expansion of isogenic cells with defined variants. Thus, beginning with target design, gene modifications can be performed within as little as 1–2 weeks, and modified clonal cell lines can be derived within 2–3 weeks [63]. Unfortunately this is presently an expensive technique which is difficult to apply to primary and poorly (or non)-proliferating cells, such as human pancreatic beta cells.

Our laboratory has been studying the role of T1D candidate genes in beta cell dysfunction and death and in the induction of local inflammation. For this purpose, we have established experimental procedures based on gene silencing and overexpression in models of human and rodent pancreatic beta cells [5–7, 11, 12]. We describe below the materials and methods used in our laboratory to select and analyze the effect of T1D candidate genes on key pathways in T1D pathogenesis, i.e., inflammation and innate immune response, apoptosis, and beta cell metabolism and function.

2   Materials

Tissue culture reagents should be prepared in an appropriate sterile environment.

2.1   Cells

1. The rat insulin-producing INS-1E cell line (kindly provided by Dr. C. Wollheim, Centre Medical Universitaire, Geneva, Switzerland [64]).

2. The human EndoC-βH1 beta cell line (kindly provided by Dr. R. Scharfmann, Centre de Recherche de l'Institut du Cerveau et de la Moelle épinière (CRICM), Paris, France).

3. FACS-purified rat beta and alpha cells from male Wistar rats, housed and used according to the guidelines of the Belgian Regulations for Animal Care. Autofluorescence-activated cell sorting is used to purify rat beta and alpha cells [65–67]. Of note, dispersed islet cells can be used in case FACS sorting facilities are not available. In this case, the procedures for culture and transfection are similar to the rat FACS-purified beta cells.

4. Human islets incubated for 2–3 min in a solution of Dispase (2.5 U/ml in Solution I) in order to disperse the islets.

1. The INS-1E cell line culture media: RPMI 1640 GlutaMAX-I, 5 % fetal bovine serum (FBS), 1 mM Na-pyruvate, 10 mM HEPES, 50 μM 2-mercaptoethanol, and 100 U/ml penicillin and 100 μg/ml streptomycin. The same medium but without antibiotics is used for transfection experiments (*see* **Note 1**).

2. The EndoC-βH1 beta cell line culture media: this cell line is cultured attached to extracellular matrix (ECM)-fibronectin-coated (100 μg/ml and 2 μg/ml, respectively as previously described [38]) (*see* **Note 2**). EndoC-βH1 cells are cultured in DMEM containing 5.6 mM glucose, 5.5 μg/ml transferrin, 6.7 ng/ml selenite, 10 mM nicotinamide, 50 μM 2-mercaptoethanol, 2 % charcoal-absorbed BSA, 100 U/ml penicillin, and 100 μg/ml streptomycin. The same medium but without antibiotics (*see* **Note 1**) and BSA (*see* **Note 3**) is used for transfection experiments. For the post-transfection recovery, 2 % FBS is added in the culture medium.

3. Rat pancreatic islet washing solution 1 (to disperse rat pancreatic islets): (5 mM glucose, 124 mM NaCl, 1 mM NaH_2PO_4, 0.71 mM $NaHCO_3$, 5.4 mM KCl, 0.8 mM H_2SO_4, and 1 mM EGTA) and incubated for 1–2 min in a solution of Dispase (5 U/ml in Solution I).

4. Dispersed human islet culture medium: Ham's F-10 containing 6.1 mM glucose, 50 μM 3-isobutyl-1-methylxanthine, 2 mM GlutaMAX, 10 % FBS, 1 % BSA, 50 U/ml penicillin, and 50 μg/ml streptomycin [68]. The transfection medium for human dispersed islets is the same used to transfect primary rat beta cells but without glucose. Dispersed islets are cultured on polylysine-coated 96-well plates.

5. FACS-purified rat beta and alpha cells culture medium: Ham's F-10 medium containing 10 mM glucose (beta cells) or 6.1 mM glucose (alpha cells), 50 μM 3-isobutyl-1-methylxanthine, 2 mM GlutaMAX, 5 % FBS (beta cells), or 10 % FBS (alpha cells), 0.5 % charcoal-absorbed BSA, 50 U/ml penicillin, and 50 μg/ml streptomycin [67, 69]. The same medium but without antibiotics (*see* **Note 1**) and BSA (*see* **Note 3**) is used for transfection experiments. Purified rat beta and alpha cells are cultured on polylysine-coated 96-well plates.

6. Phosphate buffered saline (PBS)

7. Solution of trypsin–EDTA at 0.5 mg/ml.

1. AllStars Negative Control siRNA (referred to as siCtrl) is reconstituted at 20 μM using the provided siRNA dilution buffer, aliquoted and stored at −80 °C (*see* **Note 4**). This control siRNA does not affect β-cell gene expression, function, or viability [70, 71].

2. siRNA against the selected candidate gene is reconstituted at 20 μM using the siRNA dilution buffer, aliquoted and stored at −80 °C.

3. The lipid reagent Lipofectamine RNAiMAX is used to transfect the siRNAs [70, 72].

4. Opti-MEM medium without any additive (*see* **Note 5**).

2.2.3 Recombinant Adenoviral Vectors and Infection Reagents

1. Recombinant adenovirus vector for the selected candidate gene (SIRION Biotech, Munich, Germany).

2. Recombinant adenovirus encoding Renilla luciferase (Ad-Luc) is used as negative control [73].

2.3 Analysis of Inflammatory Markers

2.3.1 mRNA Isolation, Reverse Transcriptase Reaction, and Real-Time PCR (RT-PCR)

1. Dynabeads mRNA DIRECT kit (Invitrogen).

2. Solutions for reverse transcriptase mix: 10× reaction buffer without MgCl$_2$, 50 mM MgCl$_2$ (both from Genecraft, Köln, Germany), 20 mM dNTP mix (Eurogentec, Seraing, Belgium), 2.5 μM random hexamers, 1 U/μl RNase inhibitor, and 2.5 U/μl reverse transcriptase (all from Applied Biosystems, Foster City, California, EUA).

3. iQ SYBR Green Supermix (Bio-Rad, Nazareth Eke, Belgium).

4. Specific primers for the genes of interest (Invitrogen).

5. MyiQ Single-Color Real-Time PCR Detection System (Bio-Rad).

2.3.2 Enzyme-Linked Immunosorbent Assay (ELISA)

1. Commercial ELISA kits for rat or human cytokines or chemokines

2. Microplate reader.

2.4 Analysis of Cell Viability and Mechanisms Leading to Beta Cell Death

1. A stock solution of Propidium Iodide (PI, Sigma) is prepared in PBS at 1 mg/ml, filtered and used at a final concentration of 5 μg/ml.

2. A stock solution of Hoechst 33342 (HO, Sigma) is prepared in PBS at 1 mg/ml, filtered and used at a final concentration of 5 μg/ml.

3. An inverted microscope (Zeiss, Zaventem, Belgium) with filters for excitation at 358 nM (HO) and 538 nm (PI) is used.

2.4.1 Expression of Proteins Related to Beta Cell Death

1. Solutions:
   – To prepare running gel: 30 % acrylamide–bis, 1.5 M Tris–HCl pH 8.8, 10 % SDS, 10 % ammonium persulfate, *N,N,N,N′*-tetramethyl-ethylenediamine (TEMED).
   – To prepare stacking gel: 30 % acrylamide–bis, 0.5 M Tris–HCl pH 6.8, 10 % SDS, 10 % ammonium persulfate, *N,N,N,N′*-tetramethyl-ethylenediamine (TEMED).

- Laemmli buffer: 62 mM Tris–HCl, 100 mM dithiothreitol (DTT), 10 % glycerol, 2 % SDS, 0.2 mg/ml bromophenol blue, 2 % β-mercaptoethanol. Adjust to pH 6.8.
- Running buffer: 25 mM Tris–HCl, 190 mM glycine, 0.1 % sodium dodecyl sulfate (SDS).
- Blotting buffer: 20 mM Tris–HCl, 150 mM glycine, 20 % methanol.
- Tris-buffered saline + Tween 20 (TBS-T): 20 mM Tris–HCl, 150 mM NaCl and 0.05 % Tween 20.

2. Isopropanol.

3. PageRuler prestained protein ladder

4. Supported Nitrocellulose membranes (Bio-Rad).

5. Solution 5 % BSA (fraction V) prepared in TBS-T and filtered.

6. Solution 5 % skimmed milk prepared in TBS-T.

7. Monoclonal or polyclonal primary antibodies against proteins of interest.

8. Secondary antibodies conjugated to horseradish peroxidase (different suppliers).

9. SuperSignal West Femto chemiluminescent substrate (Thermo Scientific).

10. Mini-PROTEAN Tetra Cell (Bio-Rad).

11. PowerPac HC High-Current Power Supply (Bio-Rad).

12. Mini Trans-Blot Electrophoretic Transfer Cell (Bio-Rad).

13. Molecular Imager ChemiBoc XRS$^+$ (Bio-Rad).

14. ImageLab software (Bio-Rad).

2.4.2 Nitrite Measurement

1. A stock solution of sodium nitrite ($NaNO_2$) is prepared in double distilled sterile water (ddH_2O) at 0.1 M (69 mg $NaNO_2$ in 10 ml ddH_2O) and stored at −20 °C. Protect from light (*see* **Note 6**).

2. A stock solution of N-1-napthylethylenediamine dihydrochloride (NED) is prepared in ddH_2O at 1 %. Protect from light (*see* **Note 6**).

3. A stock solution of sulfanilamide is prepared in ddH_2O plus H_3PO_4 at 10 %. 1 g of sulfanilamide is solubilized in 4.1 ml ddH_2O. Then add 5.9 ml H_3PO_4 to the aqueous solution.

4. Culture medium according to the cell type.

5. 96-well flat-bottom enzymatic assay plate (BD Falcon).

6. Microplate reader.

2.5 Analysis of Beta Cell Function: Glucose Metabolism and Insulin Release

2.5.1 Glucose Metabolism

1. Krebs-Ringer bicarbonate HEPES buffer (KRBH): 114 mM NaCl, 4.74 mM KCl, 25 mM $NaHCO_3$, 10 mM HEPES, 1.18 mM $MgSO_4$, 1.15 mM KH_2PO_4, 4.26 mM NaOH, 1.5 mM $CaCl_2$, and 0.1 % BSA. The medium pH is maintained at 7.4 (*see* **Note 7**).

2. Metabolic poison buffer: 10 ml of citrate buffer (400 mM, pH 4.9), 3 mM KCN, 5 µM antimycin A, 10 µM rotenone.

3. Radioactive D-$(U-^{14}C)$ glucose (300 mCi/mmol).

4. A stock solution of glucose monohydrate (Sigma) is prepared in KRBH at 32 mM.

5. Solution of trypsin–EDTA at 0.5 mg/ml.

6. A gas mixture of O_2 and CO_2 (95 %:5 %).

7. Solution of Hyamine hydroxide 10-×.

8. Small vial tubes made of soda-lime glass with the following dimensions: 6.75 × 35 mm × thickness 0.95 mm (flat bottom).

9. Scintillation liquid or gel.

10. Liquid scintillation counter.

2.5.2 Insulin Secretion and Content

1. Krebs-Ringer solution (KRB): 114 mM NaCl, 4.74 mM KCl, 5 mM $NaHCO_3$, 0.5 mM NaH_2PO_4, 0.5 mM $MgCl_2$, 2.54 mM $CaCl_2$, 10 mM HEPES, and 0.1 % BSA (fraction V).

2. Glucose-free complete RPMI: RPMI 1640 without glucose, 5 % heat-inactivated fetal calf serum (FCS), 10 mM HEPES, 1 mM Na-pyruvate, 50 µM 2-mercaptoethanol.

3. Stimulation media: The compositions of these media are as followed:
   - 16.7 mM glucose: 49.6 mg glucose monohydrate + 15 ml KRB.
   - 1.67 mM glucose: 500 µl glucose 16.7 mM + 4.5 ml KRB.
   - 10 mM glucose: 3.6 ml glucose 16.7 mM + 2.4 ml KRB.
   - 16.7 glucose + forskolin 10 µM: 6 ml glucose 16.7 mM + 3 µl forskolin (initial stock: 20 mM).
   - 1.67 glucose + 30 mM KCl: 600 µl glucose 16.7 mM + 180 µl KCl (initial stock: 1 M) + 5.22 ml KRB.

4. Acid ethanol solution (95 % ethanol + 5 % 12 M HCl).

5. Ultrasonic probe.

6. Rat or human insulin ELISA kit (Mercodia, Uppsala, Sweden).

3 Methods

3.1 Preparation of the Cells for Transfection or Infection

3.1.1 INS-1E Cells

1. The rat insulinoma cell line INS-1E is cultured at 37 °C. After reaching a high confluence (around 80 %), cells are washed with PBS and detached with trypsin–EDTA. Trypsin–EDTA is neutralized with INS-1E culture medium and cells are centrifuged at $1400 \times g/4$ min at room temperature. After centrifugation, the pellet is resuspended in fresh culture medium (*see* **Note 1**).

2. After counting the number of cells using a Neubauer chamber, INS-1E cells are plated at 1.0×10^4 cells/well in a 96-well plate or 1.0×10^5 cells/well in a 24-well plate depending on the type of experiment (e.g., cell viability and insulin secretion are done in 96-well plates while experiments involving RNA or protein extraction are performed in 24-well plates).

3. The cells are cultured for at least 48 h in antibiotic-free medium prior to transfection/infection.

3.1.2 Human EndoC-βH1 Cells

1. Plates are coated with extracellular matrix (ECM)-fibronectin coating medium before passage (*see* **Note 8**).

2. The human EndoC-βH1 cells are cultured at 37 °C. After reaching a high confluence (around 80 %), cells are washed with PBS and detached with trypsin–EDTA. Trypsin–EDTA is neutralized with a neutralization solution (80 % PBS + 20 % FCS). After centrifugation ($700 \times g/5$ min), the pellet is resuspended in fresh EndoC-βH1 culture medium.

3. After counting the cells using a Neubauer chamber, human EndoC-βH1 cells are plated at $4.0–7.0 \times 10^4$ cells/well in a 96-well plate.

4. The cells are cultured for at least 48 h in antibiotic-free medium prior to transfection/infection.

3.1.3 Rat Primary Beta and Alpha Cells and Human Dispersed Islets

1. FACS-purified rat beta and alpha cells and dispersed human islets are obtained as previously described [66, 67], and plated at 3.0×10^4 cells/well in 96-well plates with the appropriate medium (*see* Sect. 2.2).

2. After 48 h, the medium is exchanged by fresh medium without antibiotics and BSA. Removal of antibiotics and BSA from the culture medium is recommended for transfection of primary rat beta and alpha cells and dispersed human islets (*see* **Note 9**).

3.2 Transfection of siRNAs

This protocol is adapted for the transfection of 30 nM siRNA. For further details regarding the design of efficient siRNAs and adequate controls, *see* Ref. [71].

<table>
<tr><td>

3.2.1 INS-1E Cells

</td><td>

1. Opti-MEM and antibiotic-free INS-1E culture media are pre-warmed at 37 °C.

2. siRNAs are thawed on ice and Lipofectamine RNAiMAX lipid reagent is kept on ice.

3. siRNAs are diluted at 300 nM in Opti-MEM. From an initial stock of 20 μM, it means a 66.6× dilution.

4. The Lipofectamine RNAiMAX lipid reagent dilution differs between transfection in a 96- or a 24-well plate (*see* **Note 10**).

   – For 96-well plates: 0.2 μl/well is used (0.2 μl of Lipofectamine RNAiMAX is diluted in 9.8 μl of Opti-MEM).

   – For 24-well plates: 1 μl/well is used (1 μl of Lipofectamine RNAiMAX is diluted in 49 μl of Opti-MEM).

5. The diluted siRNA and Lipofectamine RNAiMAX are incubated separately for 5 min at room temperature and then mixed at a ratio of 1:1.

   – For 96-well plates: 10 μl of each diluted solution (20 μl final volume).

   – For 24-well plates: 50 μl of each diluted solution (100 μl final volume).

6. This new solution is gently mixed and incubated for 20 min at room temperature without agitation.

7. After 20 min, the siRNA/lipid complex is diluted 1:5 with antibiotic-free INS-1E medium.

   – For 96-well plates: 80 μl of antibiotic-free medium is added to the 20 μl of siRNA/lipid mix (100 μl final volume)

   – For 24-well plates: 400 μl of antibiotic-free medium is added to the 100 μl of siRNA/lipid mix (500 μl final volume)

8. The culture medium is removed and the transfection mix (100 μl for a 96-well and 500 μl for a 24-well) is added for overnight incubation at 37 °C (*see* **Note 11**).

9. The next day, the transfection mix is exchanged by fresh INS-1E medium for a given recovery period (24–96 h) (*see* **Note 12**).

</td></tr>
<tr><td>

3.2.2 Primary Rat Beta and Alpha Cells, Human EndoC-βH1 Cells, and Human Dispersed Islets

</td><td>

The protocol used to transfect primary rat beta and alpha cells, human EndoC-βH1 cells and human dispersed islets is similar to the one used for transfection of INS-1E cells in 96-well plates but with a different concentration of Lipofectamine RNAiMAX.

1. Primary rat beta and alpha cells, human EndoC-βH1 cells and human dispersed islets require a higher concentration of

</td></tr>
</table>

Lipofectamine RNAiMAX than INS1-E cells to be efficiently transfected (*see* **Note 10**).

– For primary rat beta and alpha cells: 0.25 μl/well is used (0.25 μl of Lipofectamine RNAiMAX is diluted in 9.75 μl of Opti-MEM).

– For human EndoC-βH1 cells and dispersed human islet cells: 0.4 μl/well is used (0.4 μl of Lipofectamine RNAi-MAX is diluted in 9.6 μl of Opti-MEM).

2. The siRNA/lipid complexes are diluted 1:5 with the antibiotic- and BSA-free transfection medium described in Sect. 2.1.

3. After overnight incubation period, the transfection medium is exchanged by culture medium specific for each cell type (*see* Sect. 2.1).

3.3 Adenoviral-Mediated Expression of Proteins

3.3.1 Use of Adenoviral Vectors

The delivery of cloned DNA into cells using adenovirus as vectors is a tool widely used in molecular biology. Two main advantages of this method are the absence of major phenotypic changes in infected cells and the fact that adenoviral infections take place in both dividing and nondividing cells at a very high efficiency without integration into the host cell's genome [74].

Generation of recombinant adenoviruses is made by deletion of the early transcription units E1 (involved in the viral replication) and E3 (involved in host immune suppression) from the adenoviral genome. These changes create a virus with insufficient viral replication ability but still able to replicate inside packaging cell lines (e.g., human kidney embryo cell 293, HEK 293) [75, 76].

To generate a recombinant adenovirus expressing a specific gene (e.g., a T1D candidate gene of interest), first the coding region of the gene must be amplified by PCR and then cloned into a shuttle vector under the control of the human cytomegalovirus (CMV) promoter (e.g., pO6-A5-CMV). Afterwards the region of interest is transferred by recombination into a plasmid containing the genome of a recombinant adenoviral vector in which the E1 and E3 transcription units have been deleted. In order to verify the presence and accuracy of the gene-open reading frame (ORF) in the resulting vector, DNA sequencing is recommended.

We describe below a protocol of infection using a recombinant adenovirus encoding Renilla luciferase (Ad-Luc) as an example. The titer of this adenovirus is 3.0×10^9 IU/ml and we use a multiplicity of infection (MOI) of 1. Of note, INS-1E cells and rat beta and alpha cells are very sensitive to toxicity secondary to adenoviral infection, and rarely survive infections with MOIs >10. Human and mouse beta cells are, however, more resistant and can be infected at higher MOIs.

1. INS-1E media (with and without antibiotics) is warmed at 37 °C.

2. Adenoviral vectors are thawed on ice.

3. The number of cells is counted in an extra well (*see* **Note 13**).
   - For 96-well plates: Wells are washed with 100 µl PBS, 20 µl trypsin are added and then the reaction is stopped with 80 µl PBS.
   - For 24-well plates: Wells are washed with 500 µl PBS, 100 µl trypsin are added and then the reaction is stopped with 400 µl PBS.

4. The volume of adenovirus needed to reach the desired MOI is calculated (*see* **Note 14**). For this purpose, the following equation is used: Volume of virus from stock (µl) = number of cells in the well × (MOI/titer of Adenovirus in PFU (plate forming units)) × 1000. In our case:
   - For 96-well plates: Volume of virus from stock (µl) = $10{,}000 \times (1/3.0 \times 10^9) \times 1000 = 0.003$ µl for one well.
   - For 24-well plates: Volume of virus from stock (µl) = $100{,}000 \times (1/3.0 \times 10^9) \times 1000 = 0.033$ µl for one well.

5. Adenovirus is diluted in antibiotic-free INS-1E medium taking into account the number of wells to be infected.

6. The culture medium is removed and the infection medium is added (100 µl for 96-well plates and 500 µl for a 24-well plates).

7. Cells are incubated during 3 h at 37 °C.

8. After the incubation period, the infection medium is removed and replaced by fresh INS-1E medium.

The protocol used to infect primary rat beta and alpha cells, human EndoC-βH1 cells and human dispersed islets is similar to the one used to infect INS-1E cells in 96-well plates. However, in these cell types the media used to dilute the adenovirus are BSA- and FBS-free (*see* **Note 15**).

3.4 Posttransfection/Postinfection Recovery Period

In experiments involving transfection of siRNAs or adenoviral-mediated expression of proteins it is important to allow an adequate recovery period after manipulating the cells for the following reasons: (a) cells need time to recover from the stressful process of transfection/infection before being challenged with another potential stressful treatment (e.g., cytokine exposure); (b) achievement of a better inhibition (by siRNAs) or overexpression (by adenoviral vectors) of target genes/proteins requires a proper recovery period. Under our experimental conditions, a recovery period of 24 h is usually sufficient to observe a clear modulation of the target protein. However, in some cases a higher recovery time (48–96 h) is necessary to observe changes in protein expression (*see* **Note 16**).

3.5 Evaluation of Efficiency

The best way to evaluate whether the expression of the gene/protein of interest has been efficiently modulated is through the measurement of its expression at mRNA and protein levels. To this aim, real-time PCR and Western blot can be used, respectively. In the case of secreted proteins (e.g., chemokines) the use of ELISA is recommended. If the target gene is a transcription factor, modulation of downstream target genes can be additionally evaluated.

In siRNA experiments, an inhibition of >50 % of the target gene expression is usually considered as efficient, but in some cases (e.g., kinases) it is necessary to reach >80 % inhibition to observe clear biological effects. Adenoviral-mediated gene overexpression based on CMV promoter usually lead to a five to tenfold increase in protein expression. In case more nuanced changes in gene/protein expression is required, an alternative is to use one approach called tetracycline-controlled transcriptional activation, in which the expression of a given gene of interest is turned on or off—Tet-On and Tet-Off, respectively—in the presence of tetracycline or doxycycline (TET-inducible adenovirus platform, SIRION Biotech). In the Tet-On system, cells are transfected with a system constituted by the tetracycline-controlled transactivator (composed by fusing the tetracycline repressor of *E. coli* with the activating domain of the herpes simplex virus VP16 protein) and the Tet-Responsive Element expression vector (formed by repeats of the tetracycline operator, a minimal CMV promoter and the gene of interest). The presence of doxycycline induces the binding of the tetracycline-controlled transactivator to the Tet-Responsive Element expression vector, thus leading to activation of gene transcription [77, 78].

3.6 Analysis of Inflammatory Markers

Proinflammatory cytokines, synthetic viral double stranded RNA (PIC; Polyinosinic–polycytidylic acid) and viral infections stimulate expression and secretion of several cytokines and chemokines in rat beta cells and in human pancreatic islets [25, 28, 34, 70, 79–82]. Our group has previously shown that the T1D candidate genes *MDA5* and *PTPN2*, and *USP18*, a member of the T1D-associated interferon regulatory factor 7-driven inflammatory network (IDIN), regulate expression of several chemokines in pancreatic beta cells [6, 10]. Among others, the following pro-inflammatory chemokines can be evaluated to assess a possible effect of a given candidate gene in pancreatic beta cell inflammation: CCL2 (MCP-1), CCL3 (MIP-1α), CCL5 (RANTES), CCL20 (MIP-3α), CXCL1 (GROα), CXCL2 (GROβ), CXCL9 (MIG), CXCL10 (IP-10), and CXCL11 (I-TAC).

In order to study the impact of T1D candidate genes in the modulation of chemokine/cytokine expression and release by beta cells real-time PCR and ELISA are respectively used.

3.6.1 mRNA Isolation, Reverse Transcriptase Reaction, and Real-Time PCR

1. After harvesting the cells with an appropriate Lysis/Binding buffer, Poly(A)$^+$ mRNA is isolated from the crude lysate using oligo (dT)$_{25}$-coated Dynabeads according to the manufacturer's instructions.

2. After mRNA isolation, cDNA is obtained by reverse transcription reaction, which is performed as follows:
   - 12 μl mRNA are added to 28 μl reverse transcriptase mix containing 1× reaction buffer without MgCl$_2$, 5 mM MgCl$_2$, 2 mM dNTP mix, 2.5 μM random hexamers, 1 U/μl RNase inhibitor, and 2.5 U/μl reverse transcriptase.
   - The mix is incubated at room temperature for 5–10 min and then at 42 °C for 1 h.
   - To stop the reaction, the samples are incubated at 99 °C for 5 min. After this step, cDNA samples can be immediately used or kept at −20 °C until assay.

3. The RT-PCR amplification reactions are performed in 20 μl containing 10 μl iQ SYBR Green Supermix, 0.5 μM of each primer (forward and reverse), 3 mM MgCl$_2$, and 2 μl template cDNA [67, 83]. The number of copies of the gene of interest is calculated by comparison with a standard curve as previously described [84] (*see* **Note 17**).

3.6.2 Enzyme-Linked Immunosorbent Assay (ELISA)

The ELISA assay is a quantitative technique based on the adsorption of some components of a reaction mixture (e.g., chemokines in the cell culture supernatant) to a stationary solid phase with special binding properties. Using the chemokines as examples, monoclonal antibodies specific for a given chemokine are pre-coated onto the microplate wells. The liquid sample, i.e., the cell culture supernatant, is added into the wells and any chemokine present binds to the antibody. After the washing steps to remove unbound substances, an enzyme-linked polyclonal antibody specific for the chemokine is added, washed and followed by color development by the product of an enzymatic reaction. The intensity of the color, which is proportional to the amount of chemokine in the sample, is measured spectrophotometrically.

Cytokines and chemokines released by pancreatic beta cells are measured by commercial ELISA kits in the cell culture supernatant according to the instructions provided by the manufacturers.

3.7 Analysis of Cell Viability and Mechanisms Leading to Beta Cell Death

Accumulating evidence suggest that a "dialogue" between invading immune cells and the target beta cells triggers beta cell death. In this context, pro-inflammatory cytokines—IL-1β, IFN-γ, and TNF-α—produced by T cells and infiltrating macrophages induce beta cell death through apoptosis [55]. Our recent findings showing that expression of several candidate genes is modulated by pro-inflammatory cytokines in pancreatic human islets and that some candidate genes play a direct role in beta cell apoptosis indicate that these genes may be important at the beta cell level as regulators of beta cell survival, and thus contribute to T1D pathogenesis [5–7, 10–12, 34].

To better understand the molecular mechanisms by which T1D candidate genes contribute to beta cell survival, in addition to measure cell viability using DNA-binding dyes, several proteins involved in different apoptotic pathways can be evaluated by RT-PCR and Western blot.

3.7.1   Cell Viability

The use of DNA-binding dyes, Hoechst 33342 (HO) and propidium iodide (PI) provides a simpler approach to evaluate cell viability in transfected/infected cells [26, 67, 85]. While HO has free passage across the plasma membrane and enters cells with preserved or damaged membranes, PI is impermeable to cells with intact membranes due to its high polarity.

1. Cells are plated and prepared for transfection/infection as described above (Sects. 3.2. and 3.3).

2. After treatment (e.g., cytokines), half of the medium is carefully removed and replaced by the same volume of culture medium containing nuclear dyes HO and PI (final concentration of 5 μg/ml).

3. After 15-min incubation, half of the staining medium is carefully removed and replaced by the same volume of fresh medium.

4. Cells are visualized and counted under an inverted microscope with filters for excitation wavelengths at:

   – 358 nm (HO, blue emission): viable and early apoptotic cells.

   – 538 nm (PI, red emission): dead cells (apoptotic + necrotic).

5. Cell viability is calculated as percentage of cell death (or apoptosis): number of dead (or apoptotic) cells/total number of cells (living + dead cells) $\times$ 100 (*see* **Note 18**).

3.7.2   Evaluation of Proteins Related to Beta Cell Death

In order to evaluate the molecular mechanisms by which a given T1D candidate gene modulates pancreatic beta cell death, we usually evaluate the intrinsic or mitochondrial pathway of cell death, the endoplasmic reticulum (ER) stress pathway and the nitric oxide (NO)-driven pathway.

3.7.3 Mitochondrial Pathway of Cell Death: Activation of Caspases 9 and 3

In mammalian cells there are two main pathways of apoptosis that lead to the activation of the final effectors, the caspases: (a) intrinsic pathway, which is initiated by events such as DNA damage, growth factor withdrawal, or cytotoxic insults. This pathway is also known as mitochondrial pathway due the key role of this organelle in the process; (b) extrinsic pathway, which is induced upon stimulation of cell surface death receptors belonging to the TNFR family (e.g., TNF-RI, Fas/CD95, and TRAIL R2/DR5) and caspase-8-cleavage/activation [86, 87].

As discussed above, pro-inflammatory cytokines induce apoptosis in beta cells. Several in vitro studies suggest that combinations of cytokines (e.g., TNF-α + IFN-γ or IL-1β + IFN-γ or IL-1β + IFN-γ + TNF-α) activate different cell death pathways [88–91]. One of the main features of cytokine-induced beta cell apoptosis is Bax translocation to the mitochondria, cytochrome c release and activation of the initiator caspase 9. Once activated, caspase 9 cleaves and activates the downstream effector caspase 3, which is essential for the execution of apoptosis. These events characterize the intrinsic pathway of apoptosis, which seems to be the main cell death pathway induced by cytokines in beta cells [90, 92, 93].

To investigate whether candidate genes are modulating the mitochondrial pathway of cell death, we measure protein levels of cleaved caspases 9 and 3 by Western blot (see below).

3.7.4 Mitochondrial Cell Death: Modulation of Bcl-2 Family Proteins

The balance between anti- and pro-apoptotic B-cell lymphoma 2 (Bcl-2) family proteins regulates the mitochondrial apoptotic pathway in beta cells [94]. This family consists of three protein groups: anti-apoptotic (Bcl-2, Bcl-XL, Mcl-1, Bcl-W, and A1), pro-apoptotic (Bax, Bak, and Bok) and BH3-only proteins. This latter group of proteins can be further subdivided in BH3-only sensitizers (DP5, Bad, Bik, Bnip3, Bmf, Noxa) and BH3-only activators (Puma, Bim, tBid) [95]. The BH3-only sensitizers are able to bind and inactivate anti-apoptotic Bcl-2 proteins, which, in turn, are displaced from their original partners (BH3-only activators). Once released, BH3-only activators are free to interact and activate pro-apoptotic Bcl-2 members, leading to formation of pores in the mitochondria, cytochrome c release and, ultimately, apoptosis [96–98].

Pro-inflammatory cytokines regulate Bcl-2 protein expression in beta cells. This regulation may occur at different levels (transcriptional or posttranscriptional) and directions (upregulation or downregulation), being dependent on the cytokine combination and time of exposure used [11, 12, 90, 99, 100]. We usually evaluate the expression of the anti-apoptotic Bcl-2, Bcl-XL, Mcl-1, A1 and of the pro-apoptotic Bax, DP5, Puma, and Bim mRNA/proteins by RT-PCR (Sect. 3.6.1) and Western blot (see below), respectively. Selection of these proteins for study is based on our previous work indicating their relevant role in cytokine-induced beta cell apoptosis [89–91, 94, 100]. As Bim has three main isoforms generated by alternative splicing, namely BimEL, BimL, and BimS [101], we evaluate all splice variants using both approaches mentioned above. Moreover, phosphorylation of Bim at serine 65 enhances its pro-apoptotic capacity [7] and thus evaluation of Bim phosphorylation should be additionally determined by Western blot.

3.7.5 Mitochondrial Cell Death: SDS Polyacrylamide Gel Electrophoresis and Western Blot

As SDS Polyacrylamide gel and Western blot techniques have been developed in detail in a previous volume (for more details, *see* Ref. [102]), in the present chapter we will briefly describe these two techniques.

1. Heat samples previously harvested in Laemmli buffer at 99 °C for 10 min and centrifuge to bring down the condensate.

2. Prepare running and stacking gels in the adequate percentage based on the molecular mass of the protein of interest.

3. Cast the running gel and gently overlay with isopropanol or water to accelerate polymerization. After polymerization, dry the isopropanol or water, add the stacking gel and insert a gel comb without introducing air bubbles.

4. When the gel is ready, load samples and protein ladder.

5. Start the electrophoresis at 80 V during 10–15 min to allow samples to stack and then continue at 100–150 V until the dye front reaches the bottom of the gel.

6. Following the electrophoresis, remove the gel carefully and transfer it to a nitrocellulose membrane previously cut to the size of the gel.

7. Perform electrophoretic transfer. We transfer gels using a Mini Trans-Blot Electrophoretic Transfer Cell (Bio-Rad) according to the instructions provided by the manufacturer. The run is carried out at 80 V during 90 min.

8. After electrophoretic transfer, remove membrane from the transfer unit and block it with 5 % skimmed milk for 1–2 h. Wash the membrane three times with TBS-T (10 min per wash).

9. Incubate overnight with primary antibody (*see* **Note 19**).

10. Following overnight incubation, remove primary antibody and wash the membrane three times with TBS-T (10 min per wash).

11. Incubate with secondary antibody for 1 h and then wash the membrane three times with TBS-T (10 min per wash).

12. Proceed with the development of the membrane. In our case we use the SuperSignal West Femto chemiluminescent

substrate. Immunoreactive bands are detected using a Molecular Imager ChemiBoc XRS⁺ and the densitometry of the bands is evaluated using Image Laboratory software.

3.7.6 ER-Stress Driven Cell Death

The correct functioning of the endoplasmic reticulum (ER) is essential for the cell. Perturbation of ER homeostasis may lead to accumulation of unfolded proteins and activation of a specific stress response known as ER stress and its consequent adaptive response, unfolded protein response (UPR) [103]. The main goal of this cellular response is to restore ER homeostasis by increasing the folding capacity and degradation of misfolded protein. When the changes mediated by the UPR do not solve the ER stress, the apoptosis pathway is activated [104].

The UPR signaling is mediated by three main ER-resident transmembrane proteins: activating transcription factor 6 (ATF6), inositol requiring ER-to-nucleus signal kinase 1α (IRE1α), and double-stranded RNA-activated kinase (PKR)-like ER kinase (PERK). These proteins are activated by accumulation of unfolded proteins in the ER lumen and transduce signals that modulate expression of key genes and proteins [105].

In beta cells, an extensive body of evidences shows that pro-inflammatory cytokines and PIC induce ER stress and apoptosis [104, 106, 107]. There is activation of several ER stress components, such as increased IRE1α expression, XBP1 splicing, PERK phosphorylation and ATF4 and CHOP induction, which triggers the mitochondrial pathway of apoptosis [106, 108]. Of note, cytokines inhibit ATF6 and BiP expression, hampering beta cell defenses against ER stress [106].

Analysis of the three UPR branches and their signaling pathways is critical to understand whether the mechanisms underlying beta cell death induced upon different stimuli (e.g., cytokines) are via ER stress activation. For this purpose we use RT-PCR and/or Western blot to measure, respectively, mRNA and/or protein levels of the three UPR mediators (ATF6, IRE1α, and PERK) as well as some key proteins related to ER stress (BiP, CHOP, spliced form of XBP1 (XBP1s) and ATF4). As PERK and eIF2α are activated by phosphorylation, the phosphorylated forms of these proteins are also evaluated by Western blot.

3.7.7 NO-Induced Cell Death

The NFκB-stimulated iNOS expression is responsible for the increase in NO production observed in beta cells exposed to cytokines [5, 109, 110]. Our group has previously shown that around 50 % of the genes modified by late (8–24 h) cytokine exposure are NO dependent, indicating a key role of this radical for the effects of cytokines and the beta cell fate [26].

Thus, in order to analyze whether a given T1D candidate gene modulates beta cell death via NO production, NO formation and secretion can be measured as describe below. It may be also of interest to determine iNOS mRNA and protein expression.

3.7.8 NO-Induced Cell Death: Nitrite Measurement

Of note, this is an adaptation of the protocol from Green et al. [111].

1. Prepare a 100 μM NaNO$_2$ solution by diluting 0.1 M NaNO$_2$ stock solution 1:1000 in the same medium used for culture. From this new solution (100 μM), prepare different dilutions (using the same culture medium) for standards: 20, 10, 5, 2.5, 1, 0.5, and 0.25 μM (*see* **Note 20**).

2. In a 96-well plate, add 200 μl of medium alone (Blank), each dilution of the standard curve and the samples. Put the Blank and standard curve samples in triplicate.

3. Prepare the Griess reagent mixing the same amounts of the sulfanilamide (stock solution at 10 %) and NED (stock solution at 1 %) solutions.

4. Add 20 μl Griess reagent in each well and incubate the plate at 65 °C for 2 min, protected from light.

5. Incubate at room temperature for 10–15 min, protected from light.

6. Measure absorbance in a microplate reader (wavelength = 546 nm) (*see* **Note 21**).

3.8 Analysis of Beta Cell Function: Glucose Metabolism and Insulin Secretion

Some candidate genes for T1D play important roles in the regulation of beta cell metabolism and function [8, 11]. The T1D susceptibility gene *GLIS3* interacts with other beta cell transcription factors (e.g., Pdx1, MafA, and NeuroD1) to increase insulin promoter activity [112]. Inhibition of *GLIS3* affects beta cell function, decreasing glucose metabolism and insulin release [11].

The protocols described below allow the measurement of glucose oxidation and insulin secretion in beta cells under different stimuli.

3.8.1 Glucose Metabolism

This protocol has been adapted from Andersson and Sandler [113].

1. Add 15 μl of D-(U-^{14}C) glucose (300 mCi/mmol; Perkin Elmer, Waltham, Massachusetts, USA) in two tubes and dry under nitrogen.

2. Add 1.6 ml of the 3.2 or 32 mM glucose solutions prepared in KRBH (one solution per tube) and mix well.

3. For each sample to be measured is necessary (work in triplicate):
   – Vials with 20 μl of the KRBH solution containing 3.2 mM glucose + D-(U-^{14}C) glucose (prepared in **step 2**).
   – Vials with 20 μl of the KRBH solution containing 32 mM glucose + D-(U-^{14}C) glucose (prepared in **step 2**).

4. Prepare the blanks (work in triplicate):
   - Blank 1.6 mM: Mix 20 μl KRBH 3.2 mM glucose + radio-active glucose with 20 μl of glucose-free KRBH.
   - Blank 16 mM: Mix 20 μl KRBH 32 mM glucose + radio-active glucose with 20 μl of glucose-free KRBH.

5. Trypsinize cells and resuspended in glucose-free KRBH solution to a density of 5.0×10^6 cells/ml.

6. Add 20 μl of the cell suspension to the tubes prepared in **step 3**. At this point, glucose concentrations in the tubes will be 1.6 or 16 mM of nonradioactive glucose and 0.19 μCi D-(U-^{14}C) glucose. Place each vial in a Packard flask, cover with rubber cap and seal.

7. Prepare the Max (tubes with the highest level of radioactivity) (work in triplicate):
   - Max 1.6 mM: add 10 μl of the solution 1.6 mM glucose prepared in the **step 2** directly in the Packard flask, cover with rubber cap and seal.
   - Max 16 mM: add 10 μl of the solution 16 mM glucose prepared in the **step 2** directly in the Packard flask, cover with rubber cap and seal.

8. Gas for 5 min with a mixture of O_2 and CO_2 (95 %:5 %). Then keep the flasks for 2 h in a water bath at 37 °C with slight agitation.

9. Inject 10 μl of metabolic poison solution into the glass vials containing samples or blanks (but not in Max). This metabolic poison solution will stop glucose oxidation.

10. Inject 250 μl of Hyamine hydroxide 10-× in the bottom of all vials (*see* **Note 22**) and incubate for 1 h at 37 °C with slight agitation.

11. Open the Packard flasks, add 6 ml of scintillation liquid or gel, mix well and close with a normal cap (*see* **Note 23**).

12. Read in a liquid scintillation counter.

3.8.2 Insulin Secretion in INS-1E Cells

1. All media are pre-warmed at 37 °C.

2. Remove the medium and replace by 500 μl glucose-free complete RPMI. Incubate for 1 h at 37 °C.

3. Remove the medium and wash cells with 500 μl KRB.

4. Add 500 μl KRB in the wells and incubate for 30 min at 37 °C.

5. Remove the medium and wash cells again with 500 μl KRB.

6. Treat cells with the different glucose stimulation media (*see* Sect. 2.5) (500 μl per well) for 30 min at 37 °C.

7. After incubation, remove 400 μl of the media, centrifuge at $300 \times g/5$ min, collect 300–350 μl supernatant and freeze at $-20\ °C$ (*insulin secretion* is measured in these samples).

8. To lyse the cells, wash with PBS, add 100 μl sterile ddH_2O, harvest the cells and collect in a tube. Do one additional wash with 100 μl sterile ddH_2O, add to the tube and freeze at $-20\ °C$ (*insulin content* is measured in these samples).

9. For insulin content measurements, sonicate samples collected in the previous step (twice for 10 s).

10. Mix 50 μl sonicated lysate with 125 μl acid ethanol solution. Freeze again at $-20\ °C$ until assay.

11. Insulin is measured by ELISA kits according to the manufacturer's instructions (Mercodia, Uppsala, Sweden).

4 Notes

1. The use of antibiotic-free medium before and during transfection is recommended to avoid cell death. This is an advice of the Lipofectamine's manufacturer (Invitrogen).

2. This medium should be prepared at 4 °C and avoid returning the ECM to room temperature (thaw it at 4 °C).

3. Previous experiments by our group showed that BSA induces aggregation of the lipid reagent, which impairs transfection efficiency.

4. Once frozen, do not thaw/freeze more than two times to avoid siRNA degradation.

5. Opti-MEM must be used free of additives (e.g., BSA and serum).

6. $NaNO_2$ and NED solutions must be protected from light. NED solution may change its color in contact with light. However, NED performance is not significantly affected by this change in color.

7. KRBH solution must be freshly prepared on the day of the experiment. If necessary, the pH can be adjusted using a gas mixture of O_2 and CO_2 (95 %:5 %) after dissolving the salts.

8. At least 1 h before starting the cell passage, plates or flasks must be coated and then placed in an incubator at 37 °C. The volume of coating medium should be added in accordance with the culture support size (e.g., ~100 μl per well for 96-well plates). Make sure that the coating medium is covering the whole surface of the culture support.

9. This step aims to remove antibiotics and BSA. As described above, these two additives may interfere with the transfection.

10. We have observed that the ratio between the number of cells and the concentration of the lipid carrier is not linear. Moreover, the cell type used affects this ratio. Thus, when starting experiments in new cell types, it is crucial to try different transfection conditions by testing different concentrations of lipid reagent and siRNA.

11. We tested shorter periods of 14–18 h without noticing significant changes in efficiency of transfection or toxicity.

12. We did not observed toxicity when using medium with antibiotics at this step.

13. The number of cells per well must be counted to calculate the MOI as accurately as possible. The counting is not necessary for cells that do not replicate, such as primary rat beta cells and dispersed human islets. In these cases, the number of cells originally plated must be used as an estimation of the number of cells per well.

14. To prepare aliquots with different MOIs, start by preparing the one with the highest MOI and make serial dilutions.

15. When setting the presently described experimental conditions, we observed that these cells were better infected in BSA- and FBS-free medium.

16. It is strongly recommended to perform a time course of 24–96 h of recovery after transfection/infection in order to establish the time point in which the maximum inhibition/overexpression of the protein is assessed. Cell viability should be evaluated in parallel to exclude nonspecific toxicity; for this purpose, non-transfected/infected cells are used as controls.

17. Standard curves are prepared for each gene of interest. To this aim a fragment of the gene is amplified by conventional PCR using specific primers. Afterwards, the PCR products are purified and serial dilutions performed in order to get the different points of the curve.

18. A minimum of 600 cells is counted in each experimental condition. Viability is evaluated by two independent observers, one of them unaware of sample identity, and the agreement between findings obtained by the two observers must be higher than 90 %.

19. Primary antibodies are diluted in 5 % BSA (fraction V) while secondary antibodies are diluted in 5 % skimmed milk in TBS-T.

20. To obtain an accurate quantitation of NO_2-levels, a nitrite standard reference curve must be prepared for each assay. This standard curve must be prepared in the same buffer used for the samples.

21. It is crucial to measure absorbance within 30 min. After this time, color may disappear.

22. Hyamine hydroxide 10-× is a $^{14}CO_2$ trapping agent that will trap all the $^{14}CO_2$ formed during glucose metabolism.

23. At this step the samples can be left in a cold room (at 4 °C) up to 24 h before reading.

References

1. Todd JA (2010) Etiology of type 1 diabetes. Immunity 32:457–467

2. Rogatsky I, Adelman K (2014) Preparing the first responders: building the inflammatory transcriptome from the ground up. Mol Cell 54:245–254

3. Chen YG, Cabrera SM, Jia S et al (2014) Molecular signatures differentiate immune states in type 1 diabetic families. Diabetes 63:3960–3973

4. Concannon P, Rich SS, Nepom GT (2009) Genetics of type 1A diabetes. N Engl J Med 360:1646–1654

5. Moore F, Colli ML, Cnop M et al (2009) PTPN2, a candidate gene for type 1 diabetes, modulates interferon-gamma-induced pancreatic beta-cell apoptosis. Diabetes 58:1283–1291

6. Colli ML, Moore F, Gurzov EN et al (2010) MDA5 and PTPN2, two candidate genes for type 1 diabetes, modify pancreatic beta-cell responses to the viral by-product double-stranded RNA. Hum Mol Genet 19:135–146

7. Santin I, Moore F, Colli ML et al (2011) PTPN2, a candidate gene for type 1 diabetes, modulates pancreatic beta-cell apoptosis via regulation of the BH3-only protein Bim. Diabetes 60:3279–3288

8. Berchtold LA, Storling ZM, Ortis F et al (2011) Huntingtin-interacting protein 14 is a type 1 diabetes candidate protein regulating insulin secretion and beta-cell apoptosis. Proc Natl Acad Sci U S A 108:E681–E688

9. Bergholdt R, Brorsson C, Palleja A et al (2012) Identification of novel type 1 diabetes candidate genes by integrating genome-wide association data, protein-protein interactions, and human pancreatic islet gene expression. Diabetes 61:954–962

10. Santin I, Eizirik DL (2013) Candidate genes for type 1 diabetes modulate pancreatic islet inflammation and beta-cell apoptosis. Diabetes Obes Metab 15(Suppl 3):71–81

11. Nogueira TC, Paula FM, Villate O et al (2013) GLIS3, a susceptibility gene for type 1 and type 2 diabetes, modulates pancreatic beta cell apoptosis via regulation of a splice variant of the BH3-only protein Bim. PLoS Genet 9:e1003532

12. Marroqui L, Santin I, Dos Santos RS et al (2014) BACH2, a candidate risk gene for type 1 diabetes, regulates apoptosis in pancreatic beta-cells via JNK1 modulation and crosstalk with the candidate gene PTPN2. Diabetes 63:2516

13. Storling J, Brorsson CA (2013) Candidate genes expressed in human islets and their role in the pathogenesis of type 1 diabetes. Curr Diab Rep 13:633–641

14. Burren OS, Guo H, Wallace C (2014) VSEAMS: a pipeline for variant set enrichment analysis using summary GWAS data identifies IKZF3, BATF and ESRRA as key transcription factors in type 1 diabetes. Bioinformatics 30:3342–3348

15. Tasan M, Musso G, Hao T et al (2015) Selecting causal genes from genome-wide association studies via functionally coherent subnetworks. Nat Methods 12:154–159

16. Evangelou M, Smyth DJ, Fortune MD et al (2014) A method for gene-based pathway analysis using genomewide association study summary statistics reveals nine new type 1 diabetes associations. Genet Epidemiol 38:661–670

17. Smink LJ, Helton EM, Healy BC et al (2005) T1DBase, a community web-based resource for type 1 diabetes research. Nucleic Acids Res 33:D544–D549

18. Wellcome Trust Case Control Consortium (2007) Genome-wide association study of 14,000 cases of seven common diseases and 3,000 shared controls. Nature 447:661–678

19. Smyth DJ, Plagnol V, Walker NM et al (2008) Shared and distinct genetic variants in type 1 diabetes and celiac disease. N Engl J Med 359:2767–2777

20. Barrett JC, Clayton DG, Concannon P et al (2009) Genome-wide association study and

meta-analysis find that over 40 loci affect risk of type 1 diabetes. Nat Genet 41:703–707

21. Bradfield JP, Qu HQ, Wang K et al (2011) A genome-wide meta-analysis of six type 1 diabetes cohorts identifies multiple associated loci. PLoS Genet 7:e1002293

22. Burren OS, Adlem EC, Achuthan P et al (2011) T1DBase: update 2011, organization and presentation of large-scale data sets for type 1 diabetes research. Nucleic Acids Res 39:D997–D1001

23. Liechti R, Csardi G, Bergmann S et al (2010) EuroDia: a beta-cell gene expression resource. Database (Oxford) 2010:baq024

24. Kutlu B, Burdick D, Baxter D et al (2009) Detailed transcriptome atlas of the pancreatic beta cell. BMC Med Genomics 2:3

25. Cardozo AK, Heimberg H, Heremans Y et al (2001) A comprehensive analysis of cytokine-induced and nuclear factor-kappa B-dependent genes in primary rat pancreatic beta-cells. J Biol Chem 276:48879–48886

26. Kutlu B, Cardozo AK, Darville MI et al (2003) Discovery of gene networks regulating cytokine-induced dysfunction and apoptosis in insulin-producing INS-1 cells. Diabetes 52:2701–2719

27. Rasschaert J, Liu D, Kutlu B et al (2003) Global profiling of double stranded RNA- and IFN-gamma-induced genes in rat pancreatic beta cells. Diabetologia 46:1641–1657

28. Ylipaasto P, Kutlu B, Rasilainen S et al (2005) Global profiling of coxsackievirus- and cytokine-induced gene expression in human pancreatic islets. Diabetologia 48:1510–1522

29. Pasquali L, Gaulton KJ, Rodriguez-Segui SA et al (2014) Pancreatic islet enhancer clusters enriched in type 2 diabetes risk-associated variants. Nat Genet 46:136–143

30. Ku GM, Kim H, Vaughn IW et al (2012) Research resource: RNA-Seq reveals unique features of the pancreatic beta-cell transcriptome. Mol Endocrinol 26:1783–1792

31. Soreq L, Guffanti A, Salomonis N et al (2014) Long non-coding RNA and alternative splicing modulations in Parkinson's leukocytes identified by RNA sequencing. PLoS Comput Biol 10:e1003517

32. Kohen R, Dobra A, Tracy JH et al (2014) Transcriptome profiling of human hippocampus dentate gyrus granule cells in mental illness. Transl Psychiatry 4:e366

33. Wu P, Zhang H, Lin W et al (2014) Discovery of novel genes and gene isoforms by integrating transcriptomic and proteomic profiling from mouse liver. J Proteome Res 13:2409

34. Eizirik DL, Sammeth M, Bouckenooghe T et al (2012) The human pancreatic islet transcriptome: expression of candidate genes for type 1 diabetes and the impact of pro-inflammatory cytokines. PLoS Genet 8: e1002552

35. Nica AC, Ongen H, Irminger JC et al (2013) Cell-type, allelic, and genetic signatures in the human pancreatic beta cell transcriptome. Genome Res 23:1554–1562

36. Bramswig NC, Everett LJ, Schug J et al (2013) Epigenomic plasticity enables human pancreatic alpha to beta cell reprogramming. J Clin Invest 123:1275–1284

37. Cnop M, Abdulkarim B, Bottu G et al (2014) RNA-sequencing identifies dysregulation of the human pancreatic islet transcriptome by the saturated fatty acid palmitate. Diabetes 63:1978

38. Ravassard P, Hazhouz Y, Pechberty S et al (2011) A genetically engineered human pancreatic beta cell line exhibiting glucose-inducible insulin secretion. J Clin Invest 121:3589–3597

39. Santin I, Moore F, Grieco FA et al (2012) USP18 is a key regulator of the interferon-driven gene network modulating pancreatic beta cell inflammation and apoptosis. Cell Death Dis 3:e419

40. Moran I, Akerman I, van de Bunt M et al (2012) Human beta cell transcriptome analysis uncovers lncRNAs that are tissue-specific, dynamically regulated, and abnormally expressed in type 2 diabetes. Cell Metab 16:435–448

41. van de Bunt M, Gaulton KJ, Parts L et al (2013) The miRNA profile of human pancreatic islets and beta-cells and relationship to type 2 diabetes pathogenesis. PLoS One 8: e55272

42. Takahashi P, Xavier DJ, Evangelista AF et al (2014) MicroRNA expression profiling and functional annotation analysis of their targets in patients with type 1 diabetes mellitus. Gene 539:213–223

43. Todd JA, Bell JI, McDevitt HO (1987) HLA-DQ beta gene contributes to susceptibility and resistance to insulin-dependent diabetes mellitus. Nature 329:599–604

44. Nejentsev S, Walker N, Riches D et al (2009) Rare variants of IFIH1, a gene implicated in antiviral responses, protect against type 1 diabetes. Science 324:387–389

45. Westra HJ, Peters MJ, Esko T et al (2013) Systematic identification of trans eQTLs as putative drivers of known disease associations. Nat Genet 45:1238–1243

46. Dina C, Meyre D, Gallina S et al (2007) Variation in FTO contributes to childhood obesity and severe adult obesity. Nat Genet 39:724–726

47. Frayling TM, Timpson NJ, Weedon MN et al (2007) A common variant in the FTO gene is associated with body mass index and predisposes to childhood and adult obesity. Science 316:889–894

48. Smemo S, Tena JJ, Kim KH et al (2014) Obesity-associated variants within FTO form long-range functional connections with IRX3. Nature 507:371–375

49. Dubois PC, Trynka G, Franke L et al (2010) Multiple common variants for celiac disease influencing immune gene expression. Nat Genet 42:295–302

50. Fehrmann RS, Jansen RC, Veldink JH et al (2011) Trans-eQTLs reveal that independent genetic variants associated with a complex phenotype converge on intermediate genes, with a major role for the HLA. PLoS Genet 7:e1002197

51. Closa A, Cordero D, Sanz-Pamplona R et al (2014) Identification of candidate susceptibility genes for colorectal cancer through eQTL analysis. Carcinogenesis 35:2039

52. GTEx Consortium (2013) The genotype-tissue expression (GTEx) project. Nat Genet 45:580–585

53. Fairfax BP, Humburg P, Makino S et al (2014) Innate immune activity conditions the effect of regulatory variants upon monocyte gene expression. Science 343:1246949

54. Gregersen PK (2014) Genetics. A genomic road map for complex human disease. Science 343:1087–1088

55. Eizirik DL, Colli ML, Ortis F (2009) The role of inflammation in insulitis and beta-cell loss in type 1 diabetes. Nat Rev Endocrinol 5:219–226

56. Hara N, Alkanani AK, Ir D, Robertson CE, Wagner BD, Frank DN, Zipris D (2013) The role of the intestinal microbiota in type 1 diabetes. Clin Immunol 146(2):112–119

57. Zipris D (2013) The interplay between the gut microbiota and the immune system in the mechanism of type 1 diabetes. Curr Opin Endocrinol Diabetes Obes 20(4):265–270

58. Seok J, Warren HS, Cuenca AG et al (2013) Genomic responses in mouse models poorly mimic human inflammatory diseases. Proc Natl Acad Sci U S A 110:3507–3512

59. Chakravarti A, Clark AG, Mootha VK (2013) Distilling pathophysiology from complex disease genetics. Cell 155:21–26

60. Heinig M, Petretto E, Wallace C et al (2010) A trans-acting locus regulates an anti-viral expression network and type 1 diabetes risk. Nature 467:460–464

61. Liu H, Beck TN, Golemis EA et al (2014) Integrating in silico resources to map a signaling network. Methods Mol Biol 1101:197–245

62. Mostafavi S, Ray D, Warde-Farley D et al (2008) GeneMANIA: a real-time multiple association network integration algorithm for predicting gene function. Genome Biol 9 (Suppl 1):S4

63. Ran FA, Hsu PD, Wright J et al (2013) Genome engineering using the CRISPR-Cas9 system. Nat Protoc 8:2281–2308

64. Merglen A, Theander S, Rubi B et al (2004) Glucose sensitivity and metabolism-secretion coupling studied during two-year continuous culture in INS-1E insulinoma cells. Endocrinology 145:667–678

65. Marroqui L, Masini M, Merino B et al (2015) Pancreatic a cells are resistant to metabolic stress-induced apoptosis in type 2 diabetes. EBioMedicine 2:378

66. Pipeleers DG, in't Veld PA, Van de Winkel M et al (1985) A new in vitro model for the study of pancreatic A and B cells. Endocrinology 117:806–816

67. Rasschaert J, Ladriere L, Urbain M et al (2005) Toll-like receptor 3 and STAT-1 contribute to double-stranded RNA+ interferon-gamma-induced apoptosis in primary pancreatic beta-cells. J Biol Chem 280:33984–33991

68. Delaney CA, Pavlovic D, Hoorens A et al (1997) Cytokines induce deoxyribonucleic acid strand breaks and apoptosis in human pancreatic islet cells. Endocrinology 138:2610–2614

69. Ling Z, Hannaert JC, Pipeleers D (1994) Effect of nutrients, hormones and serum on survival of rat islet beta cells in culture. Diabetologia 37:15–21

70. Moore F, Naamane N, Colli ML et al (2011) STAT1 is a master regulator of pancreatic {beta}-cell apoptosis and islet inflammation. J Biol Chem 286:929–941

71. Moore F, Cunha DA, Mulder H et al (2012) Use of RNA interference to investigate cytokine signal transduction in pancreatic beta cells. Methods Mol Biol 820:179–194

72. Moore F, Santin I, Nogueira TC et al (2012) The transcription factor C/EBP delta has anti-apoptotic and anti-inflammatory roles in pancreatic beta cells. PLoS One 7:e31062

73. Heimberg H, Heremans Y, Jobin C et al (2001) Inhibition of cytokine-induced NF-kappaB activation by adenovirus-mediated expression of a NF-kappaB super-repressor prevents beta-cell apoptosis. Diabetes 50:2219–2224

74. Breyer B, Jiang W, Cheng H et al (2001) Adenoviral vector-mediated gene transfer for human gene therapy. Curr Gene Ther 1:149–162

75. Graham FL, Smiley J, Russell WC et al (1977) Characteristics of a human cell line transformed by DNA from human adenovirus type 5. J Gen Virol 36:59–74

76. Fallaux FJ, Kranenburg O, Cramer SJ et al (1996) Characterization of 911: a new helper cell line for the titration and propagation of early region 1-deleted adenoviral vectors. Hum Gene Ther 7:215–222

77. Gossen M, Bujard H (1992) Tight control of gene expression in mammalian cells by tetracycline-responsive promoters. Proc Natl Acad Sci U S A 89:5547–5551

78. Gossen M, Freundlieb S, Bender G et al (1995) Transcriptional activation by tetracyclines in mammalian cells. Science 268:1766–1769

79. Chen MC, Proost P, Gysemans C et al (2001) Monocyte chemoattractant protein-1 is expressed in pancreatic islets from prediabetic NOD mice and in interleukin-1 beta-exposed human and rat islet cells. Diabetologia 44:325–332

80. Liu D, Cardozo AK, Darville MI et al (2002) Double-stranded RNA cooperates with interferon-gamma and IL-1 beta to induce both chemokine expression and nuclear factor-kappa B-dependent apoptosis in pancreatic beta-cells: potential mechanisms for viral-induced insulitis and beta-cell death in type 1 diabetes mellitus. Endocrinology 143:1225–1234

81. Cardozo AK, Proost P, Gysemans C et al (2003) IL-1beta and IFN-gamma induce the expression of diverse chemokines and IL-15 in human and rat pancreatic islet cells, and in islets from pre-diabetic NOD mice. Diabetologia 46:255–266

82. Schulte BM, Lanke KH, Piganelli JD et al (2012) Cytokine and chemokine production by human pancreatic islets upon enterovirus infection. Diabetes 61:2030–2036

83. Kharroubi I, Rasschaert J, Eizirik DL et al (2003) Expression of adiponectin receptors in pancreatic beta cells. Biochem Biophys Res Commun 312:1118–1122

84. Overbergh L, Valckx D, Waer M et al (1999) Quantification of murine cytokine mRNAs using real time quantitative reverse transcriptase PCR. Cytokine 11:305–312

85. Hoorens A, Van de Casteele M, Kloppel G et al (1996) Glucose promotes survival of rat pancreatic beta cells by activating synthesis of proteins which suppress a constitutive apoptotic program. J Clin Invest 98:1568–1574

86. Duprez L, Wirawan E, Vanden Berghe T et al (2009) Major cell death pathways at a glance. Microbes Infect 11:1050–1062

87. Fulda S, Gorman AM, Hori O et al (2010) Cellular stress responses: cell survival and cell death. Int J Cell Biol 2010:214074

88. Kim WH, Lee JW, Gao B et al (2005) Synergistic activation of JNK/SAPK induced by TNF-alpha and IFN-gamma: apoptosis of pancreatic beta-cells via the p53 and ROS pathway. Cell Signal 17:1516–1532

89. Gurzov EN, Ortis F, Cunha DA et al (2009) Signaling by IL-1beta + IFN-gamma and ER stress converge on DP5/Hrk activation: a novel mechanism for pancreatic beta-cell apoptosis. Cell Death Differ 16:1539–1550

90. Gurzov EN, Germano CM, Cunha DA et al (2010) p53 up-regulated modulator of apoptosis (PUMA) activation contributes to pancreatic beta-cell apoptosis induced by proinflammatory cytokines and endoplasmic reticulum stress. J Biol Chem 285:19910–19920

91. Allagnat F, Cunha D, Moore F et al (2011) Mcl-1 downregulation by pro-inflammatory cytokines and palmitate is an early event contributing to beta-cell apoptosis. Cell Death Differ 18:328–337

92. Holohan C, Szegezdi E, Ritter T et al (2008) Cytokine-induced beta-cell apoptosis is NO-dependent, mitochondria-mediated and inhibited by BCL-XL. J Cell Mol Med 12:591–606

93. Grunnet LG, Aikin R, Tonnesen MF et al (2009) Proinflammatory cytokines activate the intrinsic apoptotic pathway in beta-cells. Diabetes 58:1807–1815

94. Gurzov EN, Eizirik DL (2011) Bcl-2 proteins in diabetes: mitochondrial pathways of beta-cell death and dysfunction. Trends Cell Biol 21:424–431

95. Youle RJ, Strasser A (2008) The BCL-2-protein family: opposing activities that mediate cell death. Nat Rev Mol Cell Biol 9:47–59

96. Bouillet P, Strasser A (2002) BH3-only proteins - evolutionarily conserved proapoptotic Bcl-2 family members essential for initiating programmed cell death. J Cell Sci 115:1567–1574

97. Kim H, Rafiuddin-Shah M, Tu HC et al (2006) Hierarchical regulation of mitochondrion-dependent apoptosis by BCL-2 subfamilies. Nat Cell Biol 8:1348–1358

98. Shamas-Din A, Brahmbhatt H, Leber B et al (2011) BH3-only proteins: orchestrators of apoptosis. Biochim Biophys Acta 1813:508–520

99. Sarkar SA, Kutlu B, Velmurugan K et al (2009) Cytokine-mediated induction of anti-apoptotic genes that are linked to nuclear factor kappa-B (NF-kappaB) signalling in human islets and in a mouse beta cell line. Diabetologia 52:1092–1101

100. Barthson J, Germano CM, Moore F et al (2011) Cytokines tumor necrosis factor-alpha and interferon-gamma induce pancreatic beta-cell apoptosis through STAT1-mediated Bim protein activation. J Biol Chem 286:39632–39643

101. O'Connor L, Strasser A, O'Reilly LA et al (1998) Bim: a novel member of the Bcl-2-family that promotes apoptosis. EMBO J 17:384–395

102. Kurien BT, Scofield RH (2009) Protein blotting and detection: methods and protocols. Humana Press, New York, NY

103. Marciniak SJ, Ron D (2006) Endoplasmic reticulum stress signaling in disease. Physiol Rev 86:1133–1149

104. Eizirik DL, Cardozo AK, Cnop M (2008) The role for endoplasmic reticulum stress in diabetes mellitus. Endocr Rev 29:42–61

105. Cnop M, Foufelle F, Velloso LA (2012) Endoplasmic reticulum stress, obesity and diabetes. Trends Mol Med 18:59–68

106. Cardozo AK, Ortis F, Storling J et al (2005) Cytokines downregulate the sarcoendoplasmic reticulum pump Ca2+ ATPase 2b and deplete endoplasmic reticulum Ca2+, leading to induction of endoplasmic reticulum stress in pancreatic beta-cells. Diabetes 54:452–461

107. Dogusan Z, Garcia M, Flamez D et al (2008) Double-stranded RNA induces pancreatic beta-cell apoptosis by activation of the toll-like receptor 3 and interferon regulatory factor 3 pathways. Diabetes 57:1236–1245

108. Eizirik DL, Cnop M (2010) ER stress in pancreatic beta cells: the thin red line between adaptation and failure. Sci Signal 3:e7

109. Darville MI, Eizirik DL (1998) Regulation by cytokines of the inducible nitric oxide synthase promoter in insulin-producing cells. Diabetologia 41:1101–1108

110. Karlsen AE, Pavlovic D, Nielsen K et al (2000) Interferon-gamma induces interleukin-1 converting enzyme expression in pancreatic islets by an interferon regulatory factor-1-dependent mechanism. J Clin Endocrinol Metab 85:830–836

111. Green LC, Wagner DA, Glogowski J et al (1982) Analysis of nitrate, nitrite, and [15N] nitrate in biological fluids. Anal Biochem 126:131–138

112. Yang Y, Chang BH, Samson SL et al (2009) The Kruppel-like zinc finger protein Glis3 directly and indirectly activates insulin gene transcription. Nucleic Acids Res 37:2529–2538

113. Andersson A, Sandler S (1983) Viability tests of cryopreserved endocrine pancreatic cells. Cryobiology 20:161–168

Part II

Prediction

Methods in Molecular Biology (2016) 1433: 57–83
DOI 10.1007/7651_2015_292
© Springer Science+Business Media New York 2015
Published online: 13 December 2015

Islet Autoantibody Analysis: Radioimmunoassays

Rebecca Wyatt and Alistair J.K. Williams

Abstract

Type 1 diabetes (T1D) is a chronic inflammatory disease, caused by the immune mediated destruction of insulin-producing β-cells in the islets of the pancreas (Ziegler and Nepom, Immunity 32(4):468–478, 2010). Semiquantitative assays with high specificity and sensitivity for T1D are now available to detect antibodies to the four major islet autoantigens: glutamate decarboxylase (GADA) (Baekkeskov et al., Nature 347(6289):151–156, 1990), the protein tyrosine phosphatase-like proteins IA-2 (IA-2A) and IA-2β (Notkins et al., Diabetes Metab Rev 14(1):85–93, 1998), zinc transporter 8 (ZnT8A) (Wenzlau et al., Proc Natl Acad Sci U S A 104(43):17040–17045, 2007), and insulin (IAA) (Palmer, Diabetes Metab Rev 3(4):1005–1015, 1987). More than 85 % of cases of newly diagnosed or future T1D can be identified by testing for antibodies to GADA and/or IA-2A/IAA, with 98 % specificity (Bingley et al., Diabet Care 24 (2):398, 2001). Overall, radioimmunoassay (RIA) is considered the de facto gold standard format for the measurement of T1D autoantibodies (Bottazzo et al., Lancet 2(7892):1279–1283, 1974; Schlosser et al., Diabetologia 53(12):2611–2620, 2010). Here we describe current methods for autoantibody measurement using RIA. These fluid phase assays use radiolabeled ligands and immunoprecipitation to quantify autoantibodies to GAD, IA-2, ZnT8, and insulin (Bonifacio et al., J Clin Endocrinol Metab 95(7):3360–3367, 2010; Long et al., Clin Endocrinol Metab 97(2):632–637, 2012; Williams et al., J Autoimmun 10(5):473–478, 1997).

Keywords: Radioimmunoassay, Autoantibodies, Immunoprecipitation, GADA, IA-2A, ZnT8A, IAA

1 Introduction

Type 1 diabetes (T1D) is a chronic inflammatory disease, caused by the immune mediated destruction of insulin-producing β-cells in the islets of the pancreas [1]. During the past 30 years, huge advances have been made in the characterization of the autoantigens recognized by anti-islet autoantibodies. Previously, the cytoplasmic islet-cell-antibody (ICA) test was the only reliable assay for measuring humoral responses that were predictive of progression to T1D [2]. Using frozen sections of type O donor human pancreas, the degree of binding of immunoglobulins to antigens in the pancreatic islets was measured using indirect immunofluorescence. The observer scored the immunofluorescence intensity of the islets in a semiquantitative manner, following comparison with the background signal given by the acinar cells of the exocrine pancreas [3]. This assay provided a useful means for predicting T1D, and advanced our understanding of the pathophysiology of the disease. ICA measurement has been largely superseded by antibody assays specific for individual antigens, but is still utilized as a second line test in some laboratories as it offers excellent sensitivity for disease. However, the ICA test is difficult to standardize and reproduce because of variations in pancreas substrate and scoring by different observers [4].

Identification of β-cell specific antigens, some of which contribute to ICA staining, has allowed development of semiquantitative assays with high specificity and sensitivity for T1D. Reliable assays are now available to detect antibodies to the four major islet autoantigens that have so far been characterized: glutamate decarboxylase (GADA) [5], the protein tyrosine phosphatase-like proteins IA-2 (IA-2A) and IA-2β [6], zinc transporter 8 (ZnT8A) [7], and insulin itself (IAA) [8]. GAD is an enzyme which catalyzes the α-decarboxylation of L-glutamic acid to synthesize the inhibitory neurotransmitter gamma-amino butyric acid (GABA). There are two distinct isoforms: GAD67 (67 kDa) and GAD65 (65 kDa), encoded by different genes [9]. Human islets predominantly express GAD65 which has been localized to microvesicles in the β-cell [10]. IA-2 and IA-2β are membrane proteins found in the insulin secretory granules that are implicated in the insulin secretion pathway [11, 12], and there is some evidence to suggest their involvement in β-cell proliferation [13, 14]. ZnT8 is a cation transporter, also localized to the membrane of insulin secretory granules of the pancreas, which delivers the zinc ions necessary for insulin maturation and storage [15–17]. There are two main polymorphic variants of the ZnT8 carboxy-terminal domain, differing by one amino acid at residue 325 [18, 19]. Since the humoral response to ZnT8 is dependent on the genotype of an individual at this residue, both variants should be tested for when screening for type 1 diabetes [20]. A third ZnT8 variant, coding for glutamine, is uncommon in most populations, and testing for this form is not usually necessary.

More than 85 % of cases of newly diagnosed or future type 1 diabetes patients can be identified by testing for antibodies to GADA and/or IA-2A/IAA, with 98 % specificity [21]. The addition of ZnT8A to this panel has further improved disease sensitivity [22]. The risk for diabetes is higher in individuals with multiple autoantibodies compared with those positive for a single autoantibody [23, 24], and is greatest in those who develop multiple autoantibodies by the age of 2 years [25–27]. Insulin autoantibodies are often the first to appear in infants and are of high affinity in those who progress to T1D [25–28]. GADA are the second most frequent antibodies that can be detected in at-risk infants and, unlike IAA which show an inverse correlation with age [29], are also common in patients who present with disease in adulthood [30]. Autoantibodies to ZnT8 develop later than IAA and GADA

[7] and are a useful marker in lower-risk groups including older individuals, in whom disease-associated IAA are less common, as well as those with low genetic risk [31]. IA-2A also often appear later than IAA and GADA during preclinical disease; however, they are considered a highly specific marker of risk, particularly when present at high titers [25, 32].

Various methods, both commercial and in house, exist for the measurement of autoantibodies. These include fluid phase radioimmunoassay (RIA), ELISA [33], the luminescence immunoprecipitation system (LIPS) [34, 35] and, more recently, an electrochemiluminescence (ECL) assay [36]. Some of the newer assay formats may offer improved performance over RIA for measurement of some islet autoantibodies as well as avoiding the difficulties associated with radioisotopes, such as increased regulation and use of reagents with limited shelf-lives. Another important benefit of some newer assays is the potential to measure several antibodies at the same time (multiplexing), reducing the overall workload and serum requirement. While duplex assays using ^{35}S methionine and ^{3}H leucine have been used extensively for measuring islet autoantibodies, ^{3}H-labeled antigens showed reduced performance compared to those labeled with ^{35}S and are therefore not recommended [37]. Radioimmunoassay however, still offer advantages in flexibility, particularly for investigating the epitope specificity [38, 39], isotype [40, 41], and affinity of autoantibody responses [42].

In order to improve and standardize measurement of autoantibodies, international workshops and proficiency programs have been run. These initiatives began in 1985 with ICA and IAA workshops run with the support of the Immunology of Diabetes Society [43, 44]. In 2000, the Diabetes Antibody Standardization Program (DASP) was established with funding from the National Institutes of Health and this was superseded in 2012 by the Islet Autoantibody Standardization Program (IASP). The first DASP proficiency evaluation aimed to assess and improve comparability of GADA, IA-2A, and IAA measurements between laboratories through blinded assay of coded sets of control and patient sera [45]. When ZnT8 later emerged as an additional major autoantigenic target of T1D, this too was included in DASP and IASP workshops [46]. Overall, RIA has been considered the de facto gold standard format for the measurement of T1D autoantibodies [3, 47]. Efforts were made to regulate methods between different laboratories and it was deemed important to increase comparability of results within and among studies. A harmonization program was established to define thresholds and develop universal GADA and IA-2A assay protocols for National Institute of Diabetes and Digestive and Kidney Diseases (NIDDK) consortia studies [37]. However, this field is constantly evolving with the development of new sensitive and specific assays that offer improved accuracy for T1D prediction [48–52].

In this chapter, we describe current RIA methods for measuring islet autoantibodies which allow sensitive and specific detection of antibodies using small quantities of serum (5–30 μL). These fluid phase assays use either ^{35}S or ^{125}I radiolabeled ligands to bind to the antibodies and immunoprecipitation with protein A or G sepharose to isolate the antigen–antibody immunocomplexes, prior to counting in liquid scintillation or gamma counters (Fig. 1). ^{35}S-methionine labeled antigens are produced in-house for quantifying autoantibodies to GAD, IA-2, and ZnT8 using the TnT® SP6 rabbit reticulocyte lysate in vitro coupled transcription translation kits (Promega) with plasmid DNA coding for the antigen of interest [53]. This method allows ^{35}S-methionine to be incorporated into the protein, which after desalting to remove unreacted radiolabel, is used as tracer in the assays. This approach does not allow for controlled formation of disulfide bonds and therefore autoantibodies to insulin, which recognize epitopes dependent on correct disulfide

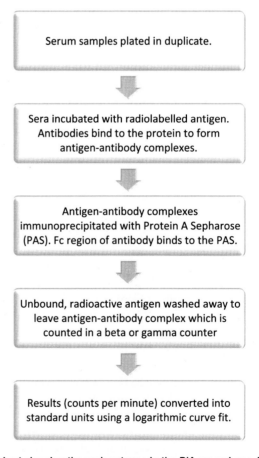

Fig. 1 A flowchart showing the major stages in the RIA procedures. The methods described below use centrifugation and aspiration to separate and wash the immunocomplexes, but this can be achieved by filtration [37, 54]

bond formation, are measured using ^{125}I labeled recombinant insulin that is sourced commercially (Perkin Elmer) [55, 56].

2 Materials

Prepare all solutions using deionized water to a resistivity of 18 M Ω cm at 25 °C and analytical grade reagents. Prepare all reagents at room temperature and store at 4 °C (unless indicated otherwise). Wear appropriate personal protective equipment, including laboratory coat and gloves. Diligently follow all waste disposal regulations when disposing of waste materials (particularly when handling radiation).

2.1 Materials Common to Both ^{35}S Methionine and ^{125}I Assay

1. Polypropylene 96-deep-well plates (Sarstedt) (*see* **Note 1**).
2. Parafilm M$^®$ (Sigma).
3. Vortex mixer (Scientific Industries, Vortex Genie 2).
4. Plate shaker (IKA-Schuttler, MTS 4).
5. Refrigerated centrifuge with deep-well plate adaptors (MSE Falcon 6/3000).
6. Vacuum pump attached to a multi-well buffer dispenser.
7. 8 prong aspirator—3.2 cm long*.
8. 8 prong aspirator—0.8 cm long*.
9. 50 % Protein A Sepharose 4 fast flow (PAS) (GE Healthcare) (*see* **Note 2**).
10. Plastic container with watertight screwcap lid.
11. Standard curve sera.
12. High positive, medium positive, low positive, and negative control sera.

Adapted from M2531 Sigma.

2.2 Materials specific to ^{35}S methionine (IA-2/GAD/ZnT8) autoantibody assays

2.2.1 Label Making

1. Methionine, L-[^{35}S]-, 1 mCi (37 MBq), Specific Activity: >1000 Ci (37.0 TBq)/mmol in 10 mM BME (Perkin Elmer, NEG009T001MC) (Store at −70 °C).
2. TnT$^®$ SP6 Quick Coupled Transcription/Translation System (Promega) (Store at −70 °C).
3. NAP5™ desalting column (packed with Sephadex™ G-25—VWR).
4. Plasmid containing the construct sequence:
   Plasmids compatible with the Promega kit should contain the required antigen construct (Table 1) and an SP6 or T7 promoter. SP6 RNA polymerase will bind to the SP6 promoter and

Table 1
Antigen constructs used to measure various islet autoantibody specificities (*see* Note 3)

Antigen	Construct	Sequence	Plasmid	Source
GAD	GAD_{65} full length	1–585	pTnT	Ake Lernmark
	GAD_{67} full length	1–594	pGEM T-easy	Vito Lampasona
	GAD_{65} (truncated)	96–585	pTnT	Vito Lampasona
IA-2	IA-2 full length	1–979	pGEM7	Ezio Bonifacio
	IA-2ic (intracellular region)	606–979	pSP64-PolyA	Vito Lampasona
	IA-2jm (juxtamembrane region)	609–631	pGEM-T	Vito Lampasona
	IA-2β (protein tyrosine phosphatase region)	723–1015	pGEM	Vito Lampasona
ZnT8	ZnT8 R (arginine at residue 325)	268–369	pTnT	Vito Lampasona
	ZnT8 W (tryptophan at residue 325)	268–369	pTnT	Vito Lampasona
	ZnT8 Q (glutamine acid at residue 325)	268–369	pTnT	Vito Lampasona

These plasmids were kindly provided by Prof. Ake Lernmark (Lund University, Malmo, Sweden); Dr Vito Lampasona, San Raffaele Hospital Scientific Institute, Milan, Italy; and Prof. Ezio Bonifacio, Dresden University of Technologies, Dresden, Germany

initiate transcription of the antigen construct to mRNA. A Kozak consensus sequence should be designed at the 5′ end of the coding sequence. When transcribed to mRNA, a favored Kozak consensus sequence such as GCCATGG, allows efficient initiation of protein translation in eukaryotes. Plasmids should also contain an antibiotic resistance gene, which will allow the plasmid to be selectively propagated. Examples of suitable plasmids include pCMVTnT and pGEM-T-easy (Promega), into which the antigen construct can be cloned.

5. Temperature-controlled water bath

2.2.2 Assay

1. Beta Scintillation counter (Topcount, Perkin Elmer).

2. White opaque 96-well Microplate (OptiPlate™—Perkin Elmer).

3. Scintillation fluid (MicroScint™ 40, Perkin Elmer).

4. Clear self-adhesive TopSeal for 96-well microplates (Perkin Elmer).

5. Tris buffered saline with Tween 20 (TBST) buffer: 20 mM Tris–HCl, pH 7.4, 150 mM NaCl, 0.15 % Tween 20. Weigh 4.844 g Tris and 17.44 g NaCl and transfer to a 2 L bottle (*see* **Note 4**). Add water to a volume of 1.8 L and mix to dissolve. Adjust pH with HCl (*see* **Note 5**). Make up to 2 L with water. Add 3 mL of Tween 20 and mix (*see* **Note 6**).

6. TBST buffer + 0.1 % BSA 20 mM Tris–HCl, pH 7.4, 150 mM NaCl, 0.15 % Tween, 0.1 % BSA. To 400 mL TBST buffer stock add 0.4 g BSA and mix to dissolve (*see* **Note 7**).

2.3 Materials Specific to ^{125}I Insulin Autoantibody Assay

1. Gamma counter.

2. ^{125}I human insulin 50 μCi (1.85 MBq), Specific Activity: >2000 Ci (74.0 TBq)/mmol (Perkin Elmer, NEX420050UC) is diluted in TBT, aliquoted and stored at −70 °C until use.

3. Soluble human insulin (e.g., Actrapid®, Novo Nordisk).

4. Tris buffer with Tween 20 (TBT): 50 mM Tris–HCl, pH, 8.0, 1 % Tween 20. Weigh 12.12 g Tris and transfer to a 2 L bottle (*see* **Note 4**). Add water to a volume of 1.8-L and mix to dissolve. Adjust pH with HCl (*see* **Note 8**). Make up to 2 L with water. Add 20 mL of Tween 20 and mix (*see* **Note 6**).

5. TBT + 1 % BSA buffer: 50 mM Tris–HCl, pH, 8.0, 1 % BSA, 1 % Tween 20. To 20 mL TBT buffer stock add 2 g BSA and mix to dissolve (*see* **Note 9**).

6. TBT/glycine buffer: 50 mM Tris–NaOH, pH, 10.6, 26 mM glycine, 1 % Tween 20. Weigh 3.03 g Tris and 9.705 g glycine and transfer to a 0.5 L bottle (*see* **Note 10**). Add 400 mL water and mix to dissolve. Adjust pH with NaOH (*see* **Note 11**). Make up to 0.5 L with water. Add 5 mL Tween 20 and mix.

7. 0.9 % sterile saline: Weigh 9 g NaCl and transfer to a 1 L bottle. Add water to volume of 1 L and mix to dissolve. Autoclave for 20 min on liquid cycle.

8. 20 % EtOH in sterile saline: 0.9 % saline, 20 % EtOH. To 40 mL 0.9 % sterile saline add 10 mL 100 % ethanol.

9. TB buffer: 50 mM Tris–HCl, pH, 8.0. Weigh 3.03 g Tris base and transfer to a 0.5 L bottle. Add water to a volume of 400 mL and mix to dissolve. Adjust pH with HCl (*see* **Note 8**). Make up to 0.5 L with water.

3 Methods

Carry out all procedures at room temperature unless otherwise specified. It is recommended to double glove when handling any radiation and wear an extremity dosimeter.

3.1 Plating Serum Samples

- Each assay should include a standard curve and quality control samples (high positive, medium positive, low positive, and negative)

- A standard curve can be split across two plates to increase the number of samples to assay. Controls should be randomly distributed and on every plate (*see* **Note 12**).

- Prior to assay, samples should be thawed and then mixed using a vortex mixer to ensure that they are homogeneous

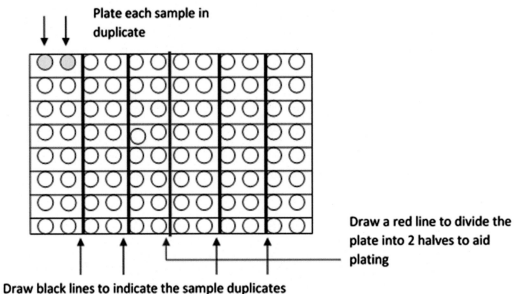

Fig. 2 Plating map for GADA/IA-2A/ZnT8A/IAA screens

- For each label, every sample should be plated in duplicate.
- 48 samples assayed in duplicate (including standard curve and control sera) will fit on GADA/IA-2A/ZnT8/IAA screen plates, while 24 samples will fit on an IAA competition plate.

3.2 GADA/IA-2A/ZnT8/IAA Screens

1. Divide a 96-deep-well plate into halves and subdivide into two well segments.

2. Place the plate on ice and pipette *2 μL* (GADA/IA-2A/Znt8A) or *5 μL* (IAA) of each sample into bottom of each pair of wells, using a clean pipette tip for every aliquot (Fig. 2) (*see* **Note 13**).

3. When all the wells contain serum, cover the plate in Parafilm M® and freeze at −20 °C or store at 4 °C if it is to be labeled the same day (*see* **Note 14**).

3.3 IAA Competition

- Low affinity and/or nonspecific insulin autoantibodies are not linked to T1D progression. Competitive displacement experiments can be conducted to increase the specificity of results.

- Serum should first be screened for IAA and then positive sera selected for competition assay.

- Samples showing increased levels of binding (>95th percentile of controls), are then re-assayed with ^{125}I insulin as normal ('hot' label) and simultaneously, an additional duplicate incubated with a label containing ^{125}I insulin and an excess concentration (8 U/mL) of cold insulin ('cold' label).

- If the autoantibodies are specific to insulin, binding of ^{125}I insulin to the antibodies will be competed by the excess cold insulin, leading to very low levels of binding. Radioactivity detectable above background levels in a sample competed with excess insulin can be considered nonspecific binding and is not directly relevant to T1D.

- The counts from the 'cold' label are therefore subtracted from the 'hot' label counts, correcting for nonspecific binding to give a final result for insulin specific antibodies.

- Each assay should include a standard curve and quality control samples (high positive, medium positive, low positive, and negative)

- A standard curve can be split across two plates to increase the number of unknown samples that can be assayed on each plate. Controls should be randomly distributed and on every plate (*see* **Note 12**).

- Samples should be plated in duplicate (four times in total—twice for 'hot' label and twice for 'cold' label). 24 samples (including standard curve and control sera) can be assayed on one plate.

- Mouse IAA may also be measured by radiobinding assay [54]. For this purpose we substitute ethanolamine-blocked protein G sepharose for glycine-blocked PAS and use a standard curve comprising dilutions of a mouse monoclonal antibody to human insulin (HUI-018, Dako) in normal mouse serum (Sigma). Human ^{125}I-insulin label is used as ligand (Perkin Elmer).

  1. Divide a 96-deep-well plate into halves and sub-divide into two well segments.

  2. Place the plate on ice and pipette 5 μL of each sample into bottom of each pair of wells for the upper half of the plate using a clean pipette tip for every aliquot (*see* **Note 13**).

  3. Repeat **step 2** for the bottom half of the plate.

  4. The first four rows and last four rows should, therefore, be identical (Fig. 3).

  5. When all the wells contain serum, cover the plate in Parafilm M$^{®}$ and freeze at -20 °C or store at 4 °C if it is to be labeled the same day (*see* **Note 14**).

3.4 Preparing Radiolabel

These instructions are adapted from those supplied for use with reagents included in Promega's TNT$^{®}$ SP6 Quick Coupled Transcription/Translation System.

1. Set temperature controlled water bath to 30 °C.

2. Thaw ^{35}S methionine stock in a fume hood ~30 min before starting the procedure (*see* **Note 15**).

3. Rapidly defrost the Master Mix and place on ice (*see* **Note 16**).

Each sample is aliquoted into 4 wells as indicated by the arrowheads

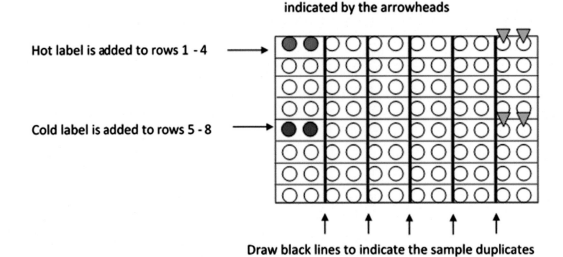

Hot label is added to rows 1 - 4

Cold label is added to rows 5 - 8

Draw black lines to indicate the sample duplicates

Fig. 3 Plating map for IAA competitions

4. Allow the nuclease free water and plasmid containing desired antigen to defrost at room temperature and place on ice.

5. To prepare the label, mix the reagents in the following order in a sterile 2.0 mL tube and keep on ice (*see* **Note 17**).

Reagent	Single (μL)	Double (μL)
Master mix	40	80
^{35}S Methionine	4	8
Plasmid (1 μg/μL)	1	2
dH$_2$O	5	10

6. Centrifuge the tube at $500 \times g$, 3 min, 4 °C and gently mix.

7. Incubate for 1 h 30 min in the water bath at 30 °C.

8. Approximately 30 min before the end of incubation, buffer exchange a NAP5 column with TBST + 0.1 % BSA buffer. Do this by allowing TBST + 0.1 % BSA buffer to drain through the column three times.

9. Remove the label from the water bath and place on ice.

10. Label eleven 2.0 mL tubes: TC, 0, 1, 2, 3, 4, 5, 6, 7, 8, 9.

11. Pipette 500 μL of TBST + 0.1 % BSA and 2 μL of label into the tube marked TC (for total count).

12. Load the full 50 μL of label onto the column. Add 50 μL TBST + 0.1 % BSA buffer to the tube (to wash) and add this

to the column. Load an additional 450 μL TBST + 0.1 % BSA buffer to the column (*see* **Notes 18–20**).

13. Collect this fraction in tube 0 (this should be clear). Add extra drops of TBST + 0.1 % BSA buffer if necessary to ensure the red fluid has reached the bottom but avoid allowing this to enter tube 0.

14. Add 300 μL TBST + 0.1 % BSA buffer and collect in tube 1. Place on ice.

15. Add 100 μL TBST + 0.1 % BSA buffer and collect in tube 2. Place on ice.

16. Add 100 μL TBST + 0.1 % BSA buffer and collect in tube 3. Place on ice.

17. Add 100 μL TBST + 0.1 % BSA buffer and collect in tube 4. Place on ice.

18. Add 100 μL TBST + 0.1 % BSA buffer and collect in tube 5.

19. Add 250 μL TBST + 0.1 % BSA buffer and collect in tube 6.

20. Add 500 μL TBST + 0.1 % BSA buffer and collect in tube 7.

21. Add 750 μL TBST + 0.1 % BSA buffer and collect in tube 8.

22. Add 1000 μL TBST + 0.1 % BSA buffer and collect in tube 9.

23. Incorporation of radiolabel into protein antigen is estimated from the elution profile by counting 2 μL from each fraction. The counts of each fraction eluted from the column is multiplied by the volume of that fraction. The sum of the volume-adjusted counts of fractions containing the labeled antigen (normally collected in tubes 1–3) is divided by the volume adjusted counts of all fractions and expressed as a percentage. This may be corrected by adjusting for the proportion of the total counts initially loaded onto the column that eluted, calculated from the TC tube.

24. Pool together fractions containing ^{35}S methionine labeled antigen, mix thoroughly and aliquot. Store at −80 °C (*see* **Note 21**).

3.5   Diluting Label

3.5.1   GADA/IA-2A/ZnT8A

1. For each plate to be assayed 2.75 mL of TBST + 0.1 % BSA buffer is required and the following volume of neat label (*see* **Notes 22** and **23**):

2. Incorporations vary from 10–20 % (ZnT8 and IA2) to 30–50 % (GADPTh65)
   Assuming that a double label is used, with a percentage incorporation of 10 %, approximately 40 μL radiolabel should be added per plate (i.e., 40 μL label to 2.75 mL of buffer).

3. Measure the total count by pipetting 25 μL of the diluted label into the well of an OptiPlate. Add 200 μL of MicroScint™-40. Cover with a TopSeal lid and mix for approximately 1 min (*see* **Note 24**).

4. Count for 1 min on a scintillation counter (*see* **Note 25**).

5. The target total count for all labels is 20,000 counts per minute (cpm) although an acceptable range is 19,000–21,000 cpm.

6. Adjust the label to meet this range by adding additional neat label or TBST + 0.1 % BSA. Repeat **steps 2** and **3** to ensure the label now reaches the required total count.

7. Add 25 μL of labeled antigen to each well of a 96-deep-well plate containing serum. Re-cover the plate with Parafilm M® and centrifuge at 500 × g, 3 min, 4 °C.

8. Mix the plates on a shaking platform for ~20 s, place on tray filled with ice and incubate for ~20 h at 4 °C (*see* **Note 26**).

3.5.2   IAA Screens

1. Remove stock ^{125}I-insulin label from the freezer and thaw at room temperature (*see* **Note 27**).

2. For each plate to be assayed measure 2.75 mL of TBT + 1 % BSA buffer and ~25 μL neat ^{125}I-insulin label (*see* **Note 23**). Mix well.

3. Pipette 25 μL of the diluted label into the bottom of a micro-tube and count on a gamma counter for ~3 min (*see* **Notes 28** and **29**).

4. The target total count is 15,000 counts per minute (cpm) although an acceptable range is 14,000–15,500 cpm.

5. Adjust the label to meet this range by adding additional neat label or TBT + 1 % BSA. Repeat **steps 3** and **4** to ensure the label now reaches the required total count.

6. Add 25 μL of diluted label to each well of a 96-deep-well plate containing serum. Re-cover the plate with Parafilm M® and centrifuge at 500 × g, 3 min, 4 °C (*see* **Note 30**).

7. Mix the plates on a shaking platform for ~20 s and incubate for 3 days at 4 °C.

3.5.3   IAA Competition

1. Remove stock ^{125}I-insulin label from the freezer and thaw at room temperature (*see* **Note 27**).

2. Label two tubes, one as IAA 'hot' label and the other as IAA 'cold' label.

3. For each plate to be assayed measure 2.5 mL of TBT + 1 % BSA buffer and ~25 μL neat ^{125}I-insulin label into the 'hot' tube. Mix well.

4. Pipette 25 μL of the diluted label into the bottom of a micro-tube and count on a gamma counter for ~3 min (*see* **Notes 28** and **29**).

5. The required total count at this stage is 16,500 counts per minute (cpm) although an acceptable range is 16,000–17,000 cpm.

6. Adjust the label to meet this range by adding additional stock label or TBT + 1 % BSA. Repeat **steps 4** and **5** to ensure the label now reaches the required total count.

7. Pour the contents of the 'hot' tube into the 'cold' tube. Coat the inside of the tube with the diluted label by swirly gently and pipette half back to the 'hot' tube.

8. Add 80 μL/mL cold insulin (100 U/mL) to the 'cold' label and mix well.

9. Add an equal volume of TBT + 1 % BSA buffer (i.e., 80 μL/mL) to the 'hot' label and mix well.

10. Pipette 25 μL of the 'hot' label into the bottom of a microtube and 25 μL of the 'cold' label into the bottom of second microtube and count on a gamma counter for ~3 min.

11. The target total count is 15,000 counts per minute (cpm) for both labels, although an acceptable range is 14,000–15,500 cpm. The 'hot' and 'cold' label should be within 500 counts of each other. Ideally, the 'cold' counts will be slightly higher than the 'hot'.

12. When both labels give satisfactory counts add 25 μL of 'hot' label to each well of the first four rows of competition plate. Add 25 μL of 'cold' label to each well of the last four rows.

13. Re-cover the plate with Parafilm M® and centrifuge at $500 \times g$, 3 min, 4 °C (*see* **Note 30**).

14. Mix the plates on a shaking platform for ~20 s and incubate for 3 days at 4 °C.

3.6 Glycine Blocking of PAS

- PAS must be glycine-blocked (GB) before use in the ^{125}I insulin assay, to reduce background counts [57].

- GB-PAS is stably preserved at 4 °C in sterile saline with 20 % ethanol.

1. Measure 10 mL of 80 % PAS slurry into a 50 mL tube (*see* **Notes 31** and **32**).

2. Fill the tube with 0.9 % sterile saline and centrifuge at $500 \times g$, 3 min, 4 °C and pour off the supernatant.

3. Repeat **step 2**.

4. Fill the tube with TBT/Glycine buffer and centrifuge at $500 \times g$, 3 min, 4 °C and pour off the supernatant.

5. Pour off the supernatant and fill tube to 50 mL with TBT/Glycine buffer.

6. Place tube onto a hybridizer at ~40 rpm and incubate overnight.

7. Centrifuge at $500 \times g$, 3 min, 4 °C and pour off the supernatant.

8. Fill the tube with TB buffer and centrifuge at $500 \times g$, 3 min, 4 °C and pour off the supernatant.

9. Fill the tube with 0.9 % sterile saline and centrifuge at $500 \times g$, 3 min, 4 °C and pour off the supernatant.

10. Repeat **step 10** two additional times.

11. Make up to a 50 % suspension with 20 % ethanol in sterile saline.

12. Wash and re-equilibrate in TBT buffer before use.

3.7 Washing/Adding PAS

3.7.1 GADA/IA-2A/ZnT8

1. Measure 2.0 mL (for ZnT8 assays) or 2.5 mL (for IA-2/GAD assays) of 50 % PAS suspension *per plate* being assayed (*see* **Notes 32** and **33**).

2. Fill the tube with TBST + 0.1 % BSA buffer. Centrifuge at $500 \times g$, 3 min, 4 °C and pour off the supernatant.

3. Repeat this wash step three additional times.

4. After the final wash step, pour off the supernatant and top up with TBST + 0.1 % BSA buffer to give a final volume of 5 mL per plate (*see* **Note 34**).

5. Add 50 μL PAS suspension into each well of the 96-deep-well plate containing serum and radiolabeled antigen that has been incubated for ~20 h.

6. Centrifuge plates at $500 \times g$, 3 min, 4 °C.

7. Incubate at 4 °C on a shaking platform (IKA-SchuttlerMTS 4) at 750 rpm for 1 h (*see* **Note 35**)

3.7.2 IAA

1. Measure 2.0 mL of 50 % GB-PAS suspension *per plate* being assayed (*see* **Notes 32** and **33**).

2. Fill the tube TBT buffer. Centrifuge at $500 \times g$, 3 min, 4 °C and pour off the supernatant.

3. Repeat this wash step three additional times.

4. After the final wash step, pour off the supernatant and top up with TBT buffer to give a final volume of 5 mL per plate (*see* **Note 34**).

5. Add 50 μL PAS suspension into each well of the 96-deep-well plate containing serum and radiolabeled antigen that has been incubated for 3 days.

6. Centrifuge plates at $500 \times g$, 3 min, 4 °C.

7. Incubate at 4 °C on a shaking platform (IKA-Schuttler MTS 4) at 750 rpm for 1 h 45 min (*see* **Note 36**)

3.8   Washing and Transferring the Assay

3.8.1   GADA/IA-2A/ZnT8A

1. After plates have incubated for 1 h with shaking, remove from the fridge and using a multi-well buffer dispenser fill wells with TBST buffer (*see* **Note 37**).

2. Centrifuge: 500 × g, 3 min, 4 °C.

3. Aspirate excess buffer with 3.2 cm aspirator, leaving a volume of ~50 μL (containing PAS) in each well (*see* **Note 38**).

4. Repeat wash step four more times.

5. After the final aspiration step keep plates at 4 °C until ready to transfer (*see* **Note 39**).

6. Add 100 μL TBST buffer to each well (*see* **Note 40**).

7. Mix the plate for ~20 s on a shaking platform (IKA-Schuttler MTS 4).

8. Using a multi-channel pipette, transfer the samples to an OptiPlate. Mix samples well with the pipette to ensure the PAS is in suspension before transferring to the OptiPlate.

9. Go back to wells to remove any residual buffer and PAS (*see* **Note 41**).

10. Centrifuge the OptiPlate: 500 × g, 3 min, 4 °C.

11. Aspirate excess buffer with 0.8 cm aspirator (*see* **Note 42**).

12. Add 20 μL MicroScint™-40 to each well. Cover the OptiPlate with a Topseal lid and shake vigorously for ~2 min (Scientific Industries, Vortex Genie 2) to form an emulsion.

13. Count on scintillation counter (*see* **Note 25**)

3.8.2   IAA and IAA Competition

1. After plates have incubated for 1 h 45 min with shaking, remove from the fridge and using a multi-well buffer dispenser fill wells with TBT buffer (*see* **Note 37**).

2. Centrifuge: 500 × g, 3 min, 4 °C.

3. Aspirate excess buffer with 3.2 cm aspirator, leaving a volume of ~50 μL (containing PAS) in each well (*see* **Note 38**).

4. Repeat wash step four more times.

5. After the final aspiration step keep plates at 4 °C until ready to transfer (*see* **Note 39**).

6. Add 100 μL TBT buffer to each well (*see* **Note 40**).

7. Mix the plate for ~20 s on the shaking platform.

8. Using a multichannel pipette, transfer the samples to micro-tubes. Mix samples well with the pipette to ensure the PAS is in suspension before transferring to the microtubes.

9. Go back to wells to remove any residual buffer and PAS (*see* **Note 41**).

10. Centrifuge microtubes: 500 × g, 3 min, 4 °C.

11. Count on gamma counter (*see* **Notes 28** and **43**).

3.9 Setting Assay Thresholds

- A population of healthy controls should be used to set a threshold of positivity. This control set should be as large as possible to allow a reliable estimate of the appropriate cut-off and ideally comprise samples from individuals with similar characteristics to those being tested.

- We normally define the positive threshold as $\geq$97.5th percentile of the controls. This should ensure a good level of sensitivity and specificity, although setting the threshold at a higher percentile (e.g., 99 %), will increase specificity with the likely trade-off being a reduction in disease sensitivity. Analysis of the receiver operator characteristics (ROC) curve will help in determining the suitability of a threshold for a particular test.

- With a small control population (<100), a threshold can be set using the mean + 3 × standard deviations. This can be useful for the initial evaluation of assays.

- Quantile-Quantile (QQ) plots can also be utilized to identify an appropriate inflection in the curve for assigning a threshold [58].

3.10 Standard Curves and Quality Controls

To ensure that assay results can be trusted rigorous quality control (QC) procedures should be instituted. This may be difficult when limited sources of QC sera are available.

- Recent-onset patients and at-risk relatives are invaluable sources of the standard and QC sera required to maintain assay performance over long periods.

- Prior to requesting a large sample from patients or relatives, a small blood sample could be taken to identify individuals with requisite antibody levels.

- Large volumes of strongly positive GADA sera are normally more easily obtained than sera with high levels of the other islet autoantibodies. This is because GADA are commonly found in older relatives and patients who are often willing and capable of donating larger blood samples. However, the epitope specificity and affinity of sera may need to be assessed when developing assays for GADA more closely associated with diabetes development. High levels of IA-2A are commonly found in children and adolescents at diagnosis of type 1 diabetes, but are less common in adult patients, limiting the volume of serum that may be available. Levels of most islet autoantibodies fall after diagnosis, but this is particularly true of ZnT8A, so blood normally needs to be collected within 2 years of diabetes onset in order to obtain moderate or high-titer ZnT8A positive sera [59]. Insulin antibody positive patients are a convenient source of standards and controls for the IAA assay, although treatment with exogenous insulin, especially if modified, may alter the epitope specificity of the antibodies. Polyclonal and monoclonal antibodies to islet antigens raised in animals may also be of use, although isotype, epitope specificity, and affinity need to be considered when choosing appropriate sera.

- Antibody negative sera from healthy donors are essential to allow for dilution of standards and provision of negative controls.

- Pooling of positive and negative sera is often necessary to achieve the large volumes required for long-term assay standardization or control and may be desirable to obtain a serum containing the average characteristics of a population.

- Ethical approval must be in place before patients, relatives and healthy donors are bled.

  1. Collect 40–50 mL blood from each volunteer into serum collection tubes and allow to clot at room temperature for 30–60 min.

  2. Centrifuge samples at 2000–3000 × g for 15 min at 4 °C.

  3. Pipette serum into a labeled sterile 30 mL universal container and gently mix.

  4. Aliquot into 2 mL tubes or freeze in bulk.

  5. Test samples in all assays.

  6. A sample with high levels of antibodies can be diluted with negative sera to generate a standard curve. Each assay should ideally have a standard curve with 6–9 points that covers the region of interest in the assay. It may be necessary to subtract counts obtained with a negative serum from those of unknowns and standards to correct for background noise in the assay.

  7. Each assay should have a negative, low, medium and high level QC sample.

  8. The low or medium QC can be made by diluting a high level sample with a negative serum or serum pool.

  9. If only small volumes of sera are available, several positive sera may have to be mixed to make one control.

  10. Standard curves and QC's should be tested multiple times in order to calculate a range of acceptable results, allowing for normal variation between assays (*see* **Note 44**).

3.11 Calculating Results

- Arbitrary units (AUs) should be assigned to each standard when setting up assays. These units should remain constant to allow for comparability of results between different assays.

- The AUs of standards should be converted to log2 to allow the generation of a logarithmic standard curve. This can be done in an Excel spreadsheet (Microsoft).

3.11.1 *GADA/IA-2A/*
ZnT8A

- The background noise of a radioimmunoassay varies between islet antigens and in our hands is usually in the range of 50–250 cpm

- To adjust for variations in background, the counts of a negative serum are subtracted from those of other samples in the GADA and IA-2A assays.

1. Calculate the mean cpm of each sample.

2. Calculate the percentage error of sample duplicates.

3. (*GADA and IA-2A only: calculate corrected cpm (ccpm) values*) Subtract the mean cpm of a known negative from the mean cpm of standard, control and unknown samples.

4. Plot the $\log_2$ AU value of each standard against its mean cpm/ccpm.

5. Fit a logarithmic trendline to the scatter graph to create a standard curve (Fig. 4).

6. Check the r^2 value of the graph. If it is below the expected value (normally >0.95), identify any bad duplicates and remove them from the standard curve (*see* **Note 45**).

7. Using the trendline equation, convert the mean cpm/ccpm of each sample to AU.

   i.e., from the curve fit in Fig. 2, the AU value of unknown (*x*) is $2^{\wedge}(1.55 \times LN(ccpm\ of\ x) - 4.9395)$

8. Check the percentage error of duplicates. Errors >30 % should be investigated and repeated if necessary (*see* **Notes 46** and **47**).

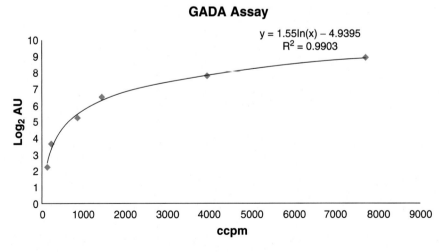

GADA Assay

$y = 1.55\ln(x) - 4.9395$
$R^2 = 0.9903$

Fig. 4 GADA standard curve

9. Check the AUs of control samples. Controls outside the expected range should be rejected. Unknown samples giving a result within the scope of the control should also be rejected.

3.11.2 IAA and IAA Competition

- The background noise of ^{125}I IAA assays, at ~5 cpm, is much lower than for ^{35}S-labeled antigens. We therefore do not correct for this, but do subtract the instrument background counts.

  1. Calculate the mean cpm of samples. In an IAA competition assay the mean cpm is calculated for 'hot' and 'cold' duplicates

  2. Calculate the percentage error of sample duplicates. In an IAA competition assay this should be calculated for 'hot' and 'cold' duplicates.

  3. (*IAA competition only: calculate corrected cpm (ccpm) values*) Subtract the 'cold' cpm of duplicate one from the 'hot' cpm of duplicate one. Repeat this for the second set of duplicates.

  4. (*IAA competition only: calculate mean ccpm values*) Subtract the mean 'cold' cpm from the mean 'hot' cpm.

  5. Plot the log$_2$ AU value of each standard against each duplicate's cpm/ccpm.

  6. Fit a logarithmic trendline to the scatter graph (Fig. 5).

  7. Check the r^2 value of the graph. If it is <0.99, identify any bad duplicates and remove them from the standard curve (*see* **Note 45**).

  8. Using the trendline equation, convert the mean cpm/ccpm of each sample to AU.

IAA Competition

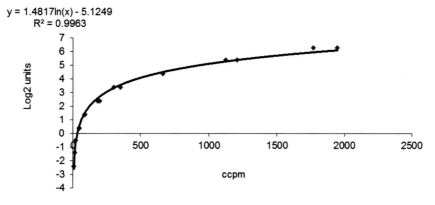

Fig. 5 IAA competition standard curve

i.e., from the curve fit in Fig. 3, the AU value of unknown (x) is $2^{\wedge}(1.4817 \times \mathrm{LN}(\text{mean ccpm of } x) - 5.1249)$

9. Check the percentage error of duplicates. Errors >30 % should be investigated and repeated if necessary (*see* **Notes 46** and **47**).

10. Check the AUs of control samples. Controls outside the expected range should be rejected. Unknown samples giving a result within the scope of the control should also be rejected.

4   Notes

1. 96-deep-well plates must have clear bottoms to aid with the 'transferring' step as it is important to be able to see any PAS at the bottom of wells.

2. If planned assay throughput is high, it is worth obtaining the PAS in bulk as this can give a large cost saving. PAS arrives at 80 % concentration and a working stock should be aliquoted and diluted to 50 % (with 20 % ethanol in sterile 0.9 % saline) as required.

3. Plasmid constructs shown to give the best performance for measuring antibodies to IA-2 and ZnT8 are based on the intracellular [59, 60] and carboxy terminal regions [61], respectively. Dimer constructs that include both ZnT8 isoforms can offer improved sensitivity when screening for ZnT8A [18]. Recently, an N-terminally truncated GAD construct encoding amino acids 96–585 was shown to identify GADA more closely associated with diabetes progression [52].

4. 2 L of buffer should be enough to assay four 96-deep-well plates.

5. 5 M HCl is convenient for adjusting the pH. Approximately 7 mL is required for a 2 L solution.

6. Buffer can be stored for up to 5 days at 4 °C. Do not use if it appears cloudy.

7. Ideally, buffer containing BSA should be used immediately, but can be stored for up to 3 days at 4 °C. Do not use if it appears cloudy. NIDDK harmonization optimized BSA concentration to 0.1 % to prevent nonspecific binding. It particularly benefits the IA-2A assay in reducing background noise [37].

8. 5 M HCl is convenient for adjusting the pH. Approximately 10.4 mL is required for 2 L.

9. TBT + 1 % BSA buffer is only used in the dilution of label and should be made fresh on the day of dilution. Research suggests ^{125}I-labeled insulin label may contain impurities, including

^{125}I-BSA. A high concentration of BSA is included in the label buffer to prevent nonspecific binding of BSA antibodies from interfering with IAA measurement [62].

10. 0.5 L of TBT/Glycine buffer will be enough to glycine block 40 mL of PAS, which should be enough to assay ~32 plates.

11. 5 M NaOH can be used to adjust the pH. Approximately 20–25 mL is required for 0.5 L.

12. As the standard curve is split across two plates, plates are paired and must be plated and assayed together.

13. Plates can be centrifuged to ensure serum is at the bottom of the wells.

14. Plates can be stored for 1–2 months at −20 °C, but samples are at risk of freeze drying. This does not seem to affect the performance of the assay, but should be avoided.

15. The ^{35}S methionine stock must be opened and used in a fume cupboard, as radioactive aerosols or gaseous degradation products may be released. Before use, mix well by pipetting slowly up and down ~30 times. Stabilized aqueous solutions suitable for in vitro translation that can be stored at 4 °C, are also available (Perkin Elmer).

16. One aliquot of Master Mix contains 200 μL. This is enough for five single or two double and one single reactions. Master Mix loses efficiency when freeze thawed. If only making single labels at a time, the 200 μL can be aliquoted to reduce freeze-thawing.

17. The volume of plasmid DNA is dependent on the concentration. 1 μL is added when the concentration = 1 mg/mL. Adjust the volume of DNA and water added to account for this.

18. The total void volume of the column is 550 μL. **Step 12** collects this volume of buffer and allows the labeled antigen to reach the bottom of the column.

19. This assumes a single label has been made, if a double is made, the total volume of the reaction is 100 μL and 100 μL of buffer is required to wash out the tube. 350 μL of buffer will then be required to reach the total void-volume of 550 μL.

20. When collecting fractions, ensure the final drop at the end of the column is caught in the tube before removing it. Always swap tubes *before* adding the buffer for the next fraction.

21. Fractions 1–4 should contain ^{35}S methionine labeled antigen, while fractions 5–9 should contain excess ^{35}S methionine which has not been incorporated into protein.

22. Neat IA-2 and GAD label can be thawed at room temperature, but should be used immediately and not left at room temperature for extended periods of time. Neat ZnT8 label should be

thawed at 4 °C and when in use should be kept on ice at all times. Diluted label can be stored temporarily at 4 °C, but should be used the same day.

23. Each serum sample is incubated with 25 μL of diluted label, but additional label is required to account for losses in tubes, from dead-space and pipetting errors. Add an additional 10 % of buffer to the final volume to ensure it does not run short. Low serum dilution promotes increased levels of binding.

24. To shake an OptiPlate vigorously, a vortex machine with an attachment to hold a plate can be used.

25. Plates should be counted in a scintillation counter and the count window optimized for detection of ^{35}S radiation. The counter needs to be normalized so that detectors yield equivalent signals for a given amount of radioactivity. To reduce background counts caused by phosphorescence, the plates should be equilibrated in the dark before counting. When diluting label to obtain high counts, it is sufficient to equilibrate for 1 min and count for 1 min. When counting a full assay, plates should be equilibrated for 10 min to minimize background counts and each well counted for 5 min or longer to minimize counting error.

26. 20 h is the standardized incubation time, although an acceptable range is 19–21 h.

27. ^{125}I-insulin label arrives in a vial as a lyophilized powder. This should be reconstituted with 2.5 mL TBT buffer using a syringe and needle to inject buffer slowly through the rubber septum into the vial. Once buffer has been added the ^{125}I-insulin is left at room temperature for 10 min, mixed and aliquoted into 2 mL screwcap microtubes. Aliquots are kept at −70 °C until use. Extensive freeze-thawing of aliquots should be avoided.

28. The multi-well gamma counter should be normalized, background-corrected, and calibrated to detect ^{125}I radiation. The counter background should be checked regularly, as even small increases can seriously affect results.

29. Readings fluctuate greatly during the first minute of counting. To ensure the counts are true, allow them to settle before taking a measurement.

30. Label should be prepared fresh each time, but any excess can be used for up to 3 days after preparation if stored at 4 °C. Counts must be adjusted on the day of use.

31. 10 mL of 80 % PAS will generate approximately 16 mL 50 % GB-PAS. This is enough to assay eight plates. Additional tubes of PAS can be blocked depending on assay throughput.

32. PAS will sink and form a layer at the bottom of its container. Ensure it has been fully resuspended (by mixing) before measuring out the stock, and both before and during addition of washed PAS to radiolabeled serum.

33. After washing, the final volume of PAS used is 5 mL per plate. Ensure the container used for washing is a suitable volume for this—i.e., a 30 mL universal tube will hold sufficient PAS for four plates (20 mL).

34. Washed PAS should be used the same day, but can be stored temporarily at 4 °C before being added to labeled plates.

35. 1 h is the standardized incubation time; plates should not be left for more than 5 min over this time before starting the next step. For GADA and IA-2A assays, optimum binding for a 1 h incubation was achieved using 2.5 mL 50 % PAS per plate [37].

36. 1 h 45 min is the optimum incubation time; plates should not be left for more than 5 min over this time before starting the next step.

37. During this process buffer should be kept in an ice water bath.

38. Aspirator should be no longer than 3.2 cm in length to ensure only excess buffer containing unbound label is removed, leaving the PAS bound to immunocomplexes at the bottom of the plate.

39. Plates should not be left for more than 2 h before being transferred.

40. Only add buffer to a plate immediately before transfer.

41. It is important that all PAS is transferred to the OptiPlate or microtubes. After the transfer is complete, wells can be checked for any residue by standing beneath the light and looking up through the bottom of the plate. If any PAS or liquid can be seen, transfer each well individually, if necessary adding extra buffer to the appropriate well.

42. Transferred plates can be stored at 4 °C for a maximum of 1 h before aspirating. Immunocomplexes will unbind from PAS in excess buffer so this step should be done promptly. After aspiration, counts should be stable, but ideally plates should be counted on the same day.

43. Each sample should be counted for 15 min to minimize counting error, with counts expressed as counts per minute (cpm).

44. Assays should include a negative, low positive, medium positive, and high positive control. These should be tested thoroughly in different assays to generate a range of values that is used to assess assay validity. The upper and lower limits that are used for judging whether an assay is acceptable can be set by calculating the mean value for the control $\pm 2 \times$ standard

deviations. These limits should be set from results of at least ten different assays and may need to be adjusted with greater experience of the assay characteristics. Coefficients of variation (SD/mean) for QC samples in autoantibody assays should ideally be below 20 %, although the CV of samples with low or negative antibody levels are often much higher than this.

45. If the required r^2 value cannot be achieved, the reliability of the assay should be questioned and the procedure examined. Samples may need to be repeated.

46. If the error is on a control sample, check if the result is within the expected range.

    • If it is in range, ignore the error.

    • If it is out of range but deleting one of the duplicates gives an acceptable result, the sample can be repeated and any samples covered by the QC considered reliable.

    • If both duplicates bring the result out of range, the control and any samples within its range, should be rejected.

47. If the error is for an unknown sample and both duplicates are below the bottom of the standard curve (i.e., negative) it can usually be ignored, unless the background is abnormally low. If one or both duplicates are above the bottom of the standard curve, the sample should be considered for repeat assay.

References

1. Ziegler AG, Nepom GT (2010) Prediction and pathogenesis in type 1 diabetes. Immunity 32(4):468–478

2. Atkinson MA, Eisenbarth GS (2001) Type 1 diabetes: new perspectives on disease pathogenesis and treatment. Lancet 358(9277):221–229

3. Bottazzo GF, Florin-Christensen A, Doniach D (1974) Islet-cell antibodies in diabetes mellitus with autoimmune polyendocrine deficiencies. Lancet 2(7892):1279–1283

4. Lernmark A, Molenaar JL, van Beers WA, Yamaguchi Y, Nagataki S, Ludvigsson J et al (1991) The Fourth International Serum Exchange Workshop to standardize cytoplasmic islet cell antibodies. The Immunology and Diabetes Workshops and Participating Laboratories. Diabetologia 34(7):534–535

5. Baekkeskov S, Aanstoot HJ, Christgau S, Reetz A, Solimena M, Cascalho M et al (1990) Identification of the 64K autoantigen in insulin-dependent diabetes as the GABA-synthesizing enzyme glutamic-acid decarboxylase. Nature 347(6289):151–156

6. Notkins AL, Lan MS, Leslie RDG (1998) IA-2 and IA-2 beta: the immune response in IDDM. Diabetes Metab Rev 14(1):85–93

7. Wenzlau JM, Juhl K, Yu L, Moua O, Sarkar SA, Gottlieb P et al (2007) The cation efflux transporter ZnT8 (Slc30A8) is a major autoantigen in human type 1 diabetes. Proc Natl Acad Sci U S A 104(43):17040–17045

8. Palmer JP (1987) Insulin autoantibodies—their role in the pathogenesis of IDDM. Diabetes Metab Rev 3(4):1005–1015

9. Erlander MG, Tillakaratne NJ, Feldblum S, Patel N, Tobin AJ (1991) Two genes encode distinct glutamate decarboxylases. Neuron 7(1):91–100

10. Kim J, Richter W, Aanstoot HJ, Shi Y, Fu Q, Rajotte R et al (1993) Differential expression of GAD65 and GAD67 in human, rat, and mouse pancreatic islets. Diabetes 42(12):1799–1808

11. Kim J, Bang H, Ko S, Jung I, Hong H, Kim-Ha J (2008) Drosophila ia2 modulates secretion of insulin-like peptide. Comp Biochem Physiol A Mol Integr Physiol 151(2):180–184

12. Kubosaki A, Gross S, Miura J, Saeki K, Zhu M, Nakamura S et al (2004) Targeted disruption of the IA-2beta gene causes glucose intolerance and impairs insulin secretion but does not prevent the development of diabetes in NOD mice. Diabetes 53(7):1684–1691

13. Torii S, Saito N, Kawano A, Hou N, Ueki K, Kulkarni RN et al (2009) Gene silencing of phogrin unveils its essential role in glucose-responsive pancreatic beta-cell growth. Diabetes 58(3):682–692

14. Mziaut H, Kersting S, Knoch KP, Fan WH, Trajkovski M, Erdmann K et al (2008) ICA512 signaling enhances pancreatic beta-cell proliferation by regulating cyclins D through STATs. Proc Natl Acad Sci U S A 105(2):674–679

15. Dodson G, Steiner D (1998) The role of assembly in insulin's biosynthesis. Curr Opin Struct Biol 8(2):189–194

16. Emdin SO, Dodson GG, Cutfield JM, Cutfield SM (1980) Role of zinc in insulin biosynthesis. Some possible zinc-insulin interactions in the pancreatic B-cell. Diabetologia 19(3):174–182

17. Chimienti F, Devergnas S, Favier A, Seve M (2004) Identification and cloning of a beta-cell-specific zinc transporter, ZnT-8, localized into insulin secretory granules. Diabetes 53 (9):2330–2337

18. Kawasaki E, Uga M, Nakamura K, Kuriya G, Satoh T, Fujishima K et al (2008) Association between anti-ZnT8 autoantibody specificities and SLC30A8 Arg325Trp variant in Japanese patients with type 1 diabetes. Diabetologia 51 (12):2299–2302

19. Wenzlau JM, Liu Y, Yu L, Moua O, Fowler KT, Rangasamy S et al (2008) A common nonsynonymous single nucleotide polymorphism in the SLC30A8 gene determines ZnT8 autoantibody specificity in type 1 diabetes. Diabetes 57(10):2693–2697

20. Wenzlau JM, Moua O, Liu Y, Eisenbarth GS, Hutton JC, Davidson HW (2008) Identification of a major humoral epitope in Slc30A8 (ZnT8). Ann N Y Acad Sci 1150:252–255

21. Bingley PJ, Bonifacio E, Ziegler AG, Schatz DA, Atkinson MA, Eisenbarth GS et al (2001) Proposed guidelines on screening for risk of type 1 diabetes. Diabetes Care 24 (2):398

22. Wenzlau JM, Moua O, Sarkar SA, Yu L, Rewers M, Eisenbarth GS et al (2008) S1C30A8 is a major target of humoral autoimmunity in type 1 diabetes and a predictive marker in prediabetes. Ann N Y Acad Sci 1150:256–259

23. Bingley PJ, Christie MR, Bonifacio E, Bonfanti R, Shattock M, Fonte MT et al (1994) Combined analysis of autoantibodies improves prediction of IDDM in islet cell antibody-positive relatives. Diabetes 43(11):1304–1310

24. Schlosser M, Strebelow M, Rjasanowski I, Kerner W, Wassmuth R, Ziegler M (2004) Prevalence of diabetes-associated autoantibodies in schoolchildren: the Karlsburg Type 1 Diabetes Risk Study. Ann N Y Acad Sci 1037:114–117

25. Ziegler AG, Hummel M, Schenker M, Bonifacio E (1999) Autoantibody appearance and risk for development of childhood diabetes in offspring of parents with type 1 diabetes: the 2-year analysis of the German BABYDIAB Study. Diabetes 48(3):460–468

26. Hummel M, Bonifacio E, Schmid S, Walter M, Knopff A, Ziegler AG (2004) Brief communication: early appearance of islet autoantibodies predicts childhood type 1 diabetes in offspring of diabetic parents. Ann Intern Med 140 (11):882–886

27. Parikka V, Nanto-Salonen K, Saarinen M, Simell T, Ilonen J, Hyoty H et al (2012) Early seroconversion and rapidly increasing autoantibody concentrations predict prepubertal manifestation of type 1 diabetes in children at genetic risk. Diabetologia 55(7):1926–1936

28. Achenbach P, Koczwara K, Knopff A, Naserke H, Ziegler AG, Bonifacio E (2004) Mature high-affinity immune responses to (pro)insulin anticipate the autoimmune cascade that leads to type 1 diabetes. J Clin Invest 114 (4):589–597

29. Vardi P, Ziegler AG, Mathews JH, Dib S, Keller RJ, Ricker AT et al (1988) Concentration of insulin autoantibodies at onset of type-i diabetes—inverse log-linear correlation with age. Diabetes Care 11(9):736–739

30. Vandewalle CL, Falorni A, Svanholm S, Lernmark A, Pipeleers DG, Gorus FK et al (1995) High diagnostic sensitivity of glutamate-decarboxylase autoantibodies in insulin-dependent diabetes-mellitus with clinical onset between age 20 and 40 years. J Clin Endocrinol Metab 80(3):846–851

31. Long AE, Gooneratne AT, Rokni S, Williams AJ, Bingley PJ (2012) The role of autoantibodies to zinc transporter 8 in prediction of type 1 diabetes in relatives: lessons from the European Nicotinamide Diabetes Intervention Trial (ENDIT) cohort. J Clin Endocrinol Metab 97(2):632–637

32. Decochez K, De Leeuw IH, Keymeulen B, Mathieu C, Rottiers R, Weets I et al (2002) IA-2 autoantibodies predict impending Type I diabetes in siblings of patients. Diabetologia 45 (12):1658–1666

33. Brooking H, Ananieva-Jordanova R, Arnold C, Amoroso M, Powell M, Betterle C et al (2003) A sensitive non-isotopic assay for GAD(65) autoantibodies. Clin Chim Acta 331 (1–2):55–59

34. Marcus P, Yan X, Bartley B, Hagopian W (2011) LIPS islet autoantibody assays in high-throughput format for DASP 2010. Diabetes Metab Res Rev 27(8):891–894

35. Burbelo PD, Ching KH, Mattson TL, Light JS, Bishop LR, Kovacs JA (2007) Rapid antibody quantification and generation of whole proteome antibody response profiles using LIPS (luciferase immunoprecipitation systems). Biochem Biophys Res Commun 352 (4):889–895

36. Yu L, Miao D, Scrimgeour L, Johnson K, Rewers M, Eisenbarth GS (2012) Distinguishing persistent insulin autoantibodies with differential risk nonradioactive bivalent proinsulin/insulin autoantibody assay. Diabetes 61(1):179–186

37. Bonifacio E, Yu L, Williams AK, Eisenbarth GS, Bingley PJ, Marcovina SM et al (2010) Harmonization of glutamic acid decarboxylase and islet antigen-2 autoantibody assays for National Institute of Diabetes and Digestive and Kidney Diseases Consortia. J Clin Endocrinol Metab 95(7):3360–3367

38. Weenink SM, Lo J, Stephenson CR, McKinney PA, Ananieva-Jordanova R, Smith BR et al (2009) Autoantibodies and associated T-cell responses to determinants within the 831–860 region of the autoantigen IA-2 in Type 1 diabetes. J Autoimmun 33(2):147–154

39. Elvers KT, Geoghegan I, Shoemark DK, Lampasona V, Bingley PJ, Williams AJ (2013) The core cysteines, (C909) of islet antigen-2 and (C945) of islet antigen-2β, are crucial to autoantibody binding in type 1 diabetes. Diabetes 62(1):214–222

40. Achenbach P, Warncke K, Reiter J, Naserke HE, Williams AJK, Bingley PJ et al (2004) Stratification of type 1 diabetes risk on the basis of islet autoantibody characteristics. Diabetes 53:384–392

41. Ronkainen MS, Hoppu S, Korhonen S, Simell S, Veijola R, Ilonen J et al (2006) Early epitope- and isotype-specific humoral immune responses to GAD65 in young children with genetic susceptibility to type 1 diabetes. Eur J Endocrinol 155(4):633–642

42. Mayr A, Schlosser M, Grober N, Kenk H, Ziegler AG, Bonifacio E et al (2007) GAD autoantibody affinity and epitope specificity identify distinct immunization profiles in children at risk for type 1 diabetes. Diabetes 56 (6):1527–1533

43. Bottazzo GF, Gleichmann H (1986) Immunology and Diabetes Workshops: report of the first international workshop on the standardisation of cytoplasmic islet cell antibodies. Diabetologia 29:125–126

44. Greenbaum CJ, Palmer JP, Kuglin B, Kolb H (1992) Insulin autoantibodies measured by radioimmunoassay methodology are more related to insulin-dependent diabetes-mellitus than those measured by enzyme-linked-immunosorbent-assay—results of the 4th international workshop on the standardization of insulin autoantibody measurement. J Clin Endocrinol Metab 74(5):1040–1044

45. Bingley PJ, Bonifacio E, Mueller PW (2003) Diabetes Antibody Standardization Program: first assay proficiency evaluation. Diabetes 52 (5):1128–1136

46. Lampasona V, Schlosser M, Mueller PW, Williams AJ, Wenzlau JM, Hutton JC et al (2011) Diabetes antibody standardization program: first proficiency evaluation of assays for autoantibodies to zinc transporter 8. Clin Chem 57 (12):1693–1702

47. Schlosser M, Mueller PW, Törn C, Bonifacio E, Bingley PJ, Laboratories P (2010) Diabetes Antibody Standardization Program: evaluation of assays for insulin autoantibodies. Diabetologia 53(12):2611–2620

48. Achenbach P, Guo L-H, Gick C, Adler K, Krause S, Bonifacio E et al (2010) A simplified method to assess affinity of insulin autoantibodies. Clin Immunol 137(3):415–421

49. Curnock RM, Reed CR, Rokni S, Broadhurst JW, Bingley PJ, Williams AJ (2012) Insulin autoantibody affinity measurement using a single concentration of unlabelled insulin competitor discriminates risk in relatives of patients with type 1 diabetes. Clin Exp Immunol 167 (1):67–72

50. Yu L, Dong F, Miao D, Fouts AR, Wenzlau JM, Steck AK (2013) Proinsulin/insulin autoantibodies measured with electrochemiluminescent assay are the earliest indicator of prediabetic islet autoimmunity. Diabetes Care 36(8):2266–2270

51. Miao D, Guyer KM, Dong F, Jiang L, Steck AK, Rewers M et al (2013) GAD65 autoantibodies detected by electrochemiluminescence assay identify high risk for type 1 diabetes. Diabetes 62(12):4174–4178

52. Williams AJK, Lampasona V, Wyatt R, Brigatti C, Gillespie KM, Bingley PJ et al (2015)

Reactivity to N-terminally truncated GAD65 (96-585) Identifies GAD autoantibodies that are more closely associated with diabetes progression in relatives of patients with type 1 diabetes. Diabetes 64:3247–3252

53. Petersen JS, Hejnaes KR, Moody A, Karlsen AE, Marshall MO, Hoiermadsen M et al (1994) Detection of GAD(65) antibodies in diabetes and other autoimmune-diseases using a simple radioligand assay. Diabetes 43 (3):459–467

54. Yu LP, Robles DT, Abiru N, Kaur P, Rewers M, Kelemen K et al (2000) Early expression of antiinsulin autoantibodies of humans and the NOD mouse: evidence for early determination of subsequent diabetes. Proc Natl Acad Sci U S A 97(4):1701–1706

55. Williams AJ, Bingley PJ, Bonifacio E, Palmer JP, Gale EA (1997) A novel micro-assay for insulin autoantibodies. J Autoimmun 10 (5):473–478

56. Naserke HE, Dozio N, Ziegler AG, Bonifacio E (1998) Comparison of a novel micro-assay for insulin autoantibodies with the conventional radiobinding assay. Diabetologia 41 (6):681–683

57. Williams AJK, Norcross AJ, Chandler KA, Bingley PJ (2006) Non-specific binding to protein A Sepharose and protein G Sepharose in insulin autoantibody assays may be reduced by pre-treatment with glycine or ethanolamine. J Immunol Methods 314(1–2):170–173

58. Graham J, Hagopian WA, Kockum I, Li LS, Sanjeevi CB, Lowe RM et al (2002) Genetic effects on age-dependent onset and islet cell autoantibody markers in type 1 diabetes. Diabetes 51(5):1346–1355

59. Wenzlau JM, Walter M, Gardner TJ, Frisch LM, Yu L, Eisenbarth GS et al (2010) Kinetics of the post-onset decline in zinc transporter 8 autoantibodies in type 1 diabetic human subjects. J Clin Endocrinol Metab 95(10):4712–4719

60. Payton MA, Hawkes CJ, Christie MR (1995) Relationship of the 37,000-M(R) and 40,000-M(R) tryptic fragments of islet antigens in insulin-dependent diabetes to the protein-tyrosine phosphatase-like molecule IA-2 (ICA512). J Clin Invest 96(3):1506–1511

61. Wenzlau JM, Frisch LM, Hutton JC, Davidson HW (2011) Mapping of conformational autoantibody epitopes in ZNT8. Diabetes Metab Res Rev 27(8):883–886

62. Williams AJK, Curnock R, Reed CR, Easton P, Rokni S, Bingley PJ (2010) Anti-BSA antibodies are a major cause of non-specific binding in insulin autoantibody radiobinding assays. J Immunol Methods 362(1-2):199–203

Methods in Molecular Biology (2016) 1433: 85–91
DOI 10.1007/7651_2015_296
© Springer Science+Business Media New York 2015
Published online: 13 December 2015

Islet Autoantibody Detection by Electrochemiluminescence (ECL) Assay

Liping Yu

Abstract

Two fundamental aspects for precisely predicting the risk of developing type 1 diabetes by islet autoantibodies are assay sensitivity and disease specificity. We have recently developed electrochemiluminescent (ECL) insulin autoantibody (IAA) and GAD65 autoantibody (GADA) assays. ECL assays are sensitive, able to identify the initiation of islet autoimmunity earlier in life among high-risk young children before clinical onset of diabetes and are more disease specific because they are able to discriminate high-affinity, high-risk diabetes specific islet autoantibodies from low-affinity, low-risk autoantibodies.

Keywords: Autoantibodies, Assay, Prediction, Diabetes

1   Introduction

Islet autoantibodies (iAbs) play an essential role in prediction of type 1 diabetes (T1D). They can appear as early as at 6 months of life and usually precede clinical diabetes by years allowing a window of opportunity to intervene. Accurate detection of the first iAb as an important biomarker for the initiation of islet autoimmunity is required to pinpoint the time of exposure to candidate environmental factors investigated by cohort studies, e.g., The Environmental Determinants of Diabetes in the Young (TEDDY) (https://teddy.epi.usf.edu/). In addition, iAbs are used extensively to stage diabetes risk and as the inclusion criteria for trials to prevent T1D. The risk of developing T1D is strongly associated with the number of iAbs among the relatives of T1D patients and the general populations. Children with two or more persistent iAbs are at high risk— 70 % will progress to diabetes in less than 10 years [1]. In contrast, children with a single persistent iAb are at a much lower risk; only 14 % have developed T1D by 10 years of follow-up. In screening for the Diabetes Prevention Trial—Type 1 (DPT-1), the majority of subjects found to be iAb positive by radioimmunoassay had single iAb to insulin (IAA) or GAD65 (GADA). In most cases, single iAb was present at low levels with low affinity. None of the 407 DPT-1 participants expressing only IAA progressed to diabetes during the initial observation period [2], casting doubt whether the presence of a single iAb is diabetes-specific. Exclusion of these "low-risk" iAb by using more specific assays would greatly enhance staging of diabetes risk for clinical trials. Recently, we developed electrochemiluminescent (ECL) IAA and GADA assays [3–5], which are more sensitive; ECL-IAA antedated the onset of islet autoimmunity by a mean of 2.3 years (range: 0.3–7.2 years) in high-risk young children followed to clinical diabetes (Diabetes Autoimmunity Study in the Young, DAISY); and more disease-specific, ECL-IAA and GADA assays are able to discriminate high affinity, high-risk iAbs in prediabetic children from those "low-risk," low-affinity signals appearing in children with single IAA or GADA.

2   Materials

1. Human proinsulin.
2. Human GAD65.
3. Biotinylation kit (ThermoScience).
4. Sulfo-TAG (MSD).
5. 96-well PCR plate (Fisher).
6. Streptavidin coated MSD plate (MSD).
7. Zeba sizing spin column (ThermoScience).
8. Blocker A buffer (MSD).
9. Reader buffer (MSD).
10. 96-well Plate Shaker (Perkin Elmer).
11. Sector 2400 or Sector 6000 (MSD).
12. Benchtop centrifuge with bucket rotary (Beckman).
13. Antigen buffer: $1 \times$ PBS with 5 % BSA.
14. PBST: $1 \times$ PBS with 0.05 % Tween 20.

3   Methods

3.1   ECL-IAA Assay

3.1.1   General Principle

Given the need for improved IAA assays and the hypothesis that the binding of insulin to solid phases obscures a key determinant for recognition by human autoantibodies as seen also for specific murine monoclonals [6], we set out to immobilize "insulin" to a solid phase in a manner that preserved critical determinants recognized by human insulin autoantibodies. We had previously failed to develop a plate capture assay for human insulin autoantibodies though we have developed such an assay (nonradioactive) for the insulin autoantibodies of the NOD mouse [7]. The two fundamental components for our recent success were utilization of proinsulin

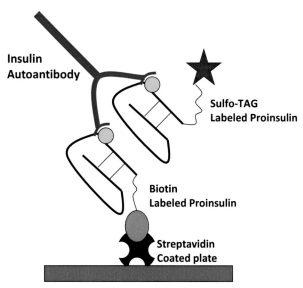

Fig. 1 Electrochemiluminescence (ECL) assay

rather than insulin and use of streptavidin rather than avidin for plate capture of biotinylated proinsulin (Fig. 1). The ruthenium assay system of the Meso Scale Discovery (MSD) Company is a third important component. In the previous studies, all human insulin autoantibodies reacted with proinsulin in fluid phase radio-assays [8].

3.1.2 Part I: Label Human Proinsulin with Biotin and Sulfo-TAG Respectively

1. Calculate the molar ratio of proinsulin with biotin or sulfo-TAG. **Note 1**. We recommend using 1:5 M ratio for proinsulin labeling. **Notes 2** and **3**.

2. Mix proinsulin with biotin or sulfo-TAG respectively, cover the reaction tubes with aluminum foil to avoid light, and incubate them at room temperature for 1 h.

3. During the 1 h incubation, prepare Zeba sizing spin column by washing the column three times with $1\times$ PBS buffer, $1000 \times g$ for 2 min each time in a centrifuge.

4. Stop the labeling reaction and purify labeled proinsulin by passing the reaction mixture through the prepared Zeba sizing spin columns with $1000 \times g$ for 2 min in a centrifuge.

5. Measure the protein concentration.

6. Aliquot the labeled proinsulin and store at $-80\,^{\circ}$C freezer.

Day 1.

1. Acid treatment of serum samples: mix 15 μL of serum with 18 μL of 0.5 M acetic acid and incubate at room temperature for 45 min.

2. To antigen buffer add 100 ng/mL of biotin and 100 ng/mL sulfo-TAG labeled proinsulin. Add 35 μL of solution per well in a new PCR plate.

3. Just before the completion of the 45 min incubation (**step 1**), add 8.3 μL of 1 M Tris pH 9.0 buffer per well into the antigen plate of **step 2** (try to add onto the side wall of each well, avoid completely mixing at this time), immediately followed by transferring 25 μL of acid treated serum from **step 1** into the well, and then mix entirely. Cover the plate with sealing foil to avoid light.

4. Shake at room temperature on low setting for 2 h.

5. Put in 4 °C refrigerator to incubate overnight (18–24 h)

Prepare the streptavidin plate.

1. Let MSD streptavidin plate(s) come to room temperature.

2. Add 150 μL of 3 % Blocker A (in PBS) per well.

3. Cover with foil.

4. Incubate in 4 °C refrigerator overnight.

Day 2.

1. Remove streptavidin plate from refrigerator. Dump out buffer and tap upside down on paper towel to dry.

2. Wash three times with 150 μL PBST.

3. Transfer 30 μL of serum/antigen mixture into the MSD streptavidin plate.

4. Cover with foil to avoid light and shake at room temperature for 1 h on low speed setting.

5. Dump out incubates and wash three times with 150 μL PBST.

6. Add 150 μL/well of 2× Read buffer (avoid any bubbles).

7. Count on MSD machine.

3.1.4 Standardization,
Quality Control, and Quality
Assurance

1. The mouse anti-human proinsulin monoclonal antibody will be used as an internal standard positive control. The monoclonal antibody should be diluted with a normal human serum and treated identically as a human serum in the assay.

2. The laboratory should keep enough volume of the positive and negative control sera for long-term use. Each of these serum samples should be aliquoted and stored at −20 °C.

3. All the assays are run in duplicate and must include positive/ negative control samples for quality control purposes and for lab index calculation. In our assay, IAA positive is any value greater than an index of 0.006.

4. The results of analysis of negative and positive controls from each assay should be plotted in a preestablished Shewart Plot with mean ±3 SD to monitor the assay drift and evaluate assay performance. Our positive control range should be within 3 SD. The negative control must be ≤0.006 for results of the assay run to be utilized.

5. Every positive sample in the first run is confirmed by repeating that sample in a different assay. If the second confirmatory assay comes out negative, a third assay is necessary. The results of two assays, which agree (e.g., +,+ or −,−), will be the final determination of positive or negative result.

3.2 ECL-GADA Assay

3.2.1 General Principle

The format of ECL-GADA assay is adapted from ECL-IAA assay except for acid treatment of serum. In a similar way to the ECL-IAA assay, a full length human GAD65 protein was set out to be immobilized to a solid phase in a manner that preserved all critical determinants recognized by human GAD65 autoantibodies.

3.2.2 Part I: Label Human GAD65 with Biotin and Sulfo-TAG Respectively

1. Calculate the molar ratio of GAD65 with biotin or sulfo-TAG. **Note 1**. We recommend using 1:20 M ratio for GAD65 labeling. **Notes 2** and **3**.

2. Mix GAD65 with biotin or sulfo-TAG respectively, cover the reaction tubes with aluminum foil to avoid light, and incubate them at room temperature for 1 h.

3. During 1 h incubation, prepare Zeba sizing spin column by washing the column three times with $1 \times$ PBS buffer, $1000 \times g$ for 2 min each time in a centrifuge.

4. Stop labeling reaction and purify labeled GAD65 by passing the reaction mixture through the prepared Zeba sizing spin columns with $1000 \times g$ for 2 min in a centrifuge.

5. Measure the protein concentration.

6. Aliquot the labeled GAD65 and store them in a −80 °C freezer.

3.2.3 Part II: Plate Capture Assay (2 Days)

Day 1.

1. Dilute 4 μL of serum with 16 μL of PBS in a PCR plate.

2. Prepare antigen buffer containing both biotin and sulfo-TAG labeled GAD65 with the concentration of 32 ng/mL for sulfo-TAG labeled GAD65 and 1000 ng/mL for biotin labeled GAD65.

3. Add 20 μL of labeled antigen solution.

4. Cover the plate with sealing foil to avoid light.

5. Shake at room temperature on low setting for 1–2 h.

6. Put in 4 °C refrigerator to incubate overnight (18–24 h)

Prepare the streptavidin plate.

1. Let MSD streptavidin plate(s) come to room temperature.

2. Add 150 µL of 3 % Blocker A (in PBS) per well.

3. Cover with foil.

4. Incubate in 4 °C refrigerator overnight.

Day 2.

1. Remove streptavidin plate from refrigerator. Dump out buffer and bang on paper towel to dry.

2. Wash three times with 150 µL PBST.

3. Transfer 30 µL of serum/antigen mixture into the MSD streptavidin plate.

4. Cover with foil to avoid light and shake at room temperature for 1 h on low speed setting.

5. Dump out incubates and Wash three times with 150 µL PBST.

6. Add 150 µL/well of 2 × Read buffer (avoid any bubbles).

7. Count on MSD machine.

3.2.4 Standardization, Quality Control, and Quality Assurance

1. The mouse anti-human GAD65 monoclonal antibody will be used as an internal standard positive control. The monoclonal antibody should be diluted with a normal human serum and treated identically as a human serum in the assay.

2. The laboratory should keep enough volume of the positive and negative control sera for long-term use. Each of these serum samples should be aliquoted and stored in −20 °C.

3. All the assays are run in duplicate and must include positive/negative control samples for quality control purpose and for lab index calculation. In our assay, GAA positive is any value greater than an index of 0.023.

4. The results of analysis of negative and positive controls from each assay should be plotted in a preestablished Shewart Plot with mean ±3 SD to monitor the assay drift and evaluate assay performance. Our positive control range should be within 3SD. The negative control must be ≤0.023 for results of the assay run to be utilized.

5. Every positive sample in the first run is confirmed by repeating that sample in a different assay. If the second confirmatory assay comes out negative, a third assay is necessary. The results of two assays, which agree (e.g., +,+ or −,−), will be the final determination of positive or negative result.

4   Notes

1. Biotin and sulfo-TAG powder should be dissolved just before labeling procedure every time.

2. The proinsulin and the GAD65 must be stored at $-80\ ^{\circ}C$ and thawed on ice before use.

3. The protein in either tris or glycine buffer systems should be exchanged to $2\times$ PBS buffer with pH 7.9 by sizing spin column.

Acknowledgement

This study was supported by NIH grant DK32083, Diabetes Autoimmunity Study in the Young Grant DK32493, the Immune Tolerance Network AI15416, the Juvenile Diabetes Research Foundation Grant 11-2005-15, DERC Clinical Core DK57516. We thank Dr. Tom Thomas of Vanderbilt for the gift of an insulin monoclonal antibody. We thank Eli Lilly Company for supplying proinsulin and thank MSD Company for helpful discussions during assay development.

References

1. Ziegler AG, Rewers M, Simell O et al (2013) Seroconversion to multiple islet autoantibodies and risk of progression to diabetes in children. JAMA 309:2473–2479

2. Orban T, Sosenko JM, Cuthbertson D et al (2009) Diabetes Prevention Trial-1 Study Group. Pancreatic islet autoantibodies as predictors of type 1 diabetes in the diabetes prevention trial-type 1 (DPT-1. Diabetes Care 32:2269–2274

3. Yu L, Miao D, Scrimgeour L, Johnson K, Rewers M, Eisenbarth GS (2012) Distinguishing persistent insulin autoantibodies with differential risk: non-radioactive bivalent proinsulin/insulin autoantibody assay. Diabetes 61:179–186

4. Miao D, Guyer KM, Dong F, Jiang L, Steck AK, Rewers M, Eisenbarth GS, Yu L (2013) GAD65 autoantibodies detected by electrochemiluminescence (ECL) assay identify high risk for type 1 diabetes. Diabetes 62:4174–8

5. Yu L, Dong F, Fuse M, Miao D, Fouts A, Hutton JC, Steck AK (2013) Proinsulin/Insulin Autoantibodies Measured with Electrochemiluminescent Assay Are the Earliest Indicator of Prediabetic Islet Autoimmunity. Diabetes Care 36:2266–2270

6. Rojas M, Hulbert C, Thomas JW (2001) Anergy and not clonal ignorance determines the fate of B cells that recognize a physiological autoantigen. J Immunol 166:3194–3200

7. Babaya N, Liu E, Miao D, Li M, Yu L, Eisenbarth GS (2009) Murine high specificity/sensitivity competitive europium insulin autoantibody assay. Diabetes Technol Ther 11:227–233

8. Castano L, Ziegler AG, Ziegler R, Shoelson S, Eisenbarth GS (1993) Characterization of insulin autoantibodies in relatives of patients with type 1 diabetes. Diabetes 42:1202–1209

Methods in Molecular Biology (2016) 1433: 93–102
DOI 10.1007/7651_2016_330
© Springer Science+Business Media New York 2016
Published online: 16 April 2016

Detection of C-Peptide in Urine as a Measure of Ongoing Beta Cell Function

T.J. McDonald and M.H. Perry

Abstract

C-peptide is a protein secreted by the pancreatic beta cells in equimolar quantities with insulin, following the cleavage of proinsulin into insulin. Measurement of C-peptide is used as a surrogate marker of endogenous insulin secretory capacity. Assessing C-peptide levels can be useful in classifying the subtype of diabetes as well as assessing potential treatment choices in the management of diabetes.

Standard measures of C-peptide involve blood samples collected either fasted or, most often, after a fixed stimulus (such as oral glucose, mixed meal, or IV glucagon). Despite the established clinical utility of blood C-peptide measurement, its widespread use is limited. In many instances this is due to perceived practical restrictions associated with sample collection.

Urine C-peptide measurement is an attractive noninvasive alternative to blood measures of beta-cell function. Urine C-peptide creatinine ratio measured in a single post stimulated sample has been shown to be a robust, reproducible measure of endogenous C-peptide which is stable for three days at room temperature when collected in boric acid. Modern high sensitivity immunoassay technologies have facilitated measurement of C-peptide down to single picomolar concentrations.

Keywords: C-peptide, Urine, UCPCR, Endogenous, Insulin

1 Introduction

Insulin secretory capacity can be assessed by measuring C-peptide, a protein secreted by the pancreas in equimolar quantities with insulin, following the enzymatic cleavage of proinsulin into insulin (Fig. 1). C-peptide is an attractive surrogate marker of insulin secretion because it can be used in patients administered exogenous insulin therapy and it is more reflective of insulin secretion than insulin itself, due to the variable clearance of insulin by the liver before it reaches the peripheral circulation [1]. C-peptide metabolism largely occurs in the kidney through glomerular filtration and uptake by tubular cells from peritubular capillaries, with 5–10 % normally excreted in the urine. This is in contrast to insulin, 50 % of which is metabolized and extracted in the liver. The total quantity of C-peptide excreted in the urine per day represents 5 % of pancreatic secretion, compared to only 0.1 % of secreted insulin [2].

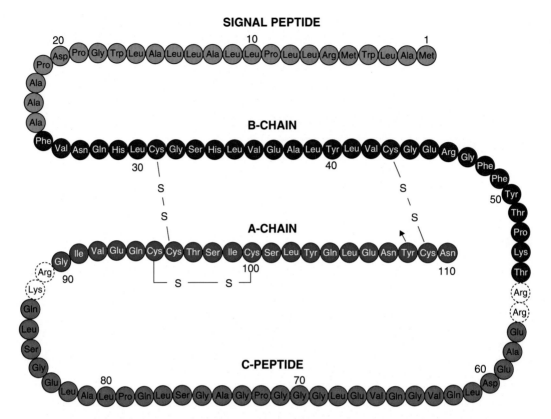

Fig. 1 Diagram of pro-insulin demonstrating the A chain and B chain of insulin and the connecting peptide (C-peptide), figure adapted from Stoy *et al* 2007[5]. During the biosynthesis of insulin, the C peptide promotes proper protein folding and disulfide bonds between the A and B chains

Consequently, despite equimolar secretion, C-peptide has a longer half-life of approximately 30 min compared with only 6 min for insulin and occurs in the blood in concentrations up to five times higher than insulin [3, 4].

1.1 C-Peptide and Differentiating Subtypes of Diabetes

The assessment of endogenous C-peptide production in diabetes can be useful in classifying the subtype of diabetes. In a patient with young onset diabetes persistent C-peptide production may reflect the honeymoon period of a patient with Type 1 diabetes, but also enduring C-peptide can be a feature of other types of diabetes including Type 2 diabetes where levels are typically high.

There is extensive evidence that C-peptide can be used to differentiate between the classification of Type 1 diabetes and Type 2 diabetes [5–9]. C-peptide is a good candidate biomarker to differentiate patients with Maturity Onset Diabetes of the Young (MODY) from Type 1 diabetes. Type 1 diabetes is characterized by autoimmune destruction of beta-cells and ultimately results in absolute insulin deficiency, usually within 5 years of diagnosis [10]. In patients with a genetic diagnosis of MODY, beta cell function is typically maintained with time despite a reduction in insulin secretion and progressive hyperglycemia [11]. The ISPAD Clinical Practise Consensus Guidelines "Compendium on the diagnosis and management of monogenic diabetes in children and adolescents" published in 2009 [12] and best practice guidelines for the molecular genetic diagnosis of maturity-onset diabetes of the young, published in 2008 [13] both acknowledge that persistent C-peptide production can be used to identify patients likely to harbor a mutation in a MODY gene. Both guidelines suggest after the initial honeymoon period in patients with suspected Type 1 diabetes (typically >5 years) a blood C-peptide level of >200 pmol/L suggests a possible diagnosis of MODY. These guidelines are supported by recent reports which have shown that a urine C-peptide can be used to discriminate Type 1 diabetes from MODY with a high sensitivity and specificity [14, 15]

1.2 Measurement of C-Peptide

1.2.1 Blood C-Peptide

Standard measures of C-peptide involve blood samples collected either fasted or, most often, after a fixed stimulus (such as oral glucose, mixed meal or IV glucagon) [1, 16, 17]. Despite the established clinical utility of blood C-peptide measurement, its widespread use is limited. In many instances this is due to practical restrictions associated with sample collection. Both insulin and C-peptide concentrations are widely considered to be unstable and require specific pre-analytical handling procedures with many laboratory providers stipulating the need for blood samples to be rapidly centrifuged after collection on ice and the serum or plasma immediately frozen following separation from cells [1, 18]. However, these strict handling protocols may not be necessary as there is evidence that C-peptide is stable for up to 24 h on whole blood if collected into EDTA preservative [19].

1.2.2 Urine C-Peptide

There is a body of evidence that suggests 24 h urine C-peptide (UCP) levels provide an accurate means of assessing beta cell-secretory capacity and correlate with both fasting and stimulated serum insulin and C-peptide [2, 7, 20, 21].

However, the cumbersome nature and difficulties in obtaining an accurate and complete 24 h urine collection has limited the utility of this test [22–24]. Urine samples collected over a period of 4 h and single post stimulated UCP have been shown to be significantly correlated with serum insulin and serum C-peptide in nondiabetic subjects [25], and in insulin treated diabetic patients [6]. Correcting for creatinine adjusts urine C-peptide concentration for variation in urine concentration and enables the use of "spot" urine samples in place of 24-h urine collection. Urinary C-peptide when collected in boric acid is stable for up to 3 days at room temperature making it possible to be collected remotely from the processing laboratory and even posted into the laboratory by patients [26]. Urine C-peptide creatinine ratio correlates strongly with blood C-peptide and has been validated in nondiabetics, type 1 diabetes, type 2 diabetes, adults and children [14, 15, 27–34].

1.2.3 C-Peptide Assays   Early assays for C–peptide were radioimmunoassays and suffered from poor analytical sensitivity and specificity and are time-consuming to perform (and therefore expensive) [1, 35, 36]. The introduction of non-isotopic assays (chemiluminescence, fluorescence, etc.) that utilize monoclonal antibodies has improved analytical sensitivity, specificity, reproducibility and reduced assay costs (to approximately £10). These technologies are also amenable to automation, allowing these assays to be incorporated into large high throughput clinical analysers.

2 Materials

- Urine collected into boric acid container. For interpretation of beta-cell reserve, the sample should be a stimulated sample, collected 2 h after the biggest meal of the day (see Methods section on patient preparation).
- Roche Modular Analytics E170 immunoassay analyzer (Roche Diagnostics, Mannheim, Germany).
- C-peptide calibration material: Roche C-peptide Calibration Set (Roche Diagnostics, Mannheim, Germany).
- C-peptide quality control material: Elecsys Precicontrol Multi-analyte (Roche Diagnostics, Mannheim, Germany).
- Roche C-peptide chemiluminescence assay kit (Roche Diagnostics, Mannheim, Germany).
- Diluent used for C-peptide assay (Equine serum albumin): Roche Diluent MultiAssay (Roche Diagnostics, Mannheim, Germany).
- Roche Creatinine modified Jaffe assay kit (Roche Diagnostics, Mannheim, Germany).
- Creatinine calibration material: Roche Calibrator for automated systems (Roche Diagnostics, Mannheim, Germany).
- Creatinine QC material: Randox assayed urine control level 2 and 3 (Randox UK, Co. Antrim, UK).

3 Methods

Patient preparation: To assess the maximum endogenous C-peptide response, a stimulated urine sample should be collected. Patients should be briefed with the following information before sample collection:

1. Pass urine just before the biggest meal of the day and discard.

2. Eat the meal as usual with a glass or more of water.

3. Do not eat anything else for the next 2 h unless you have a hypoglycemic episode, in which case you should do this test on another day. You can drink water freely throughout the duration of the collection.

4. 2 h after this meal, collect urine and send for analysis.

3.1 Sample Storage and Stability

Urine C-peptide concentration is unchanged at room temperature for 24 h and at 4 °C for 72 h even in the absence of preservative.

Urine C-peptide collected in boric acid is stable at room temperature for 72 h. Urine C-peptide remained stable after seven freeze–thaw cycles but decreased with freezer storage time and dropped to 82–84 % of baseline by 90 days at −20 °C [26].

3.2 Roche Assay Used to Measure C-Peptide in Urine

C-peptide analysis is performed on the high throughput Roche Modular Analytics E170 immunoassay analyzer (Roche Diagnostics, Mannheim, Germany). The assay is a heterogeneous sandwich immunoassay. Urine is diluted 10-fold with Multi Assay Diluent. One biotinylated anti-C-peptide specific monoclonal mouse antibody and a second monoclonal antibody to C-peptide labeled with a ruthenium complex, react with C-peptide in 20 µL of diluted urine sample to form an antigen–antibody–antigen sandwich complex. Separation is achieved via interaction of biotin and streptavidin attachment to paramagnetic microparticles (solid phase). The detection system employs electrochemiluminescence with ruthenium trisbipyridyl as the label.

Electrochemiluminescence occurs at 620 nm and readings are taken by the photomultiplier tube. The intensity of light signal is proportional to the concentration of C-peptide in the serum. All urine samples were pre-diluted 1:10 with equine serum albumin (diluent multianalyte, Roche Diagnostics, Mannheim, Germany) (*see* **Note 2**). The analysis time is 18 min per specimen. The E170 is able to analyze 170 samples per hour.

The assay was calibrated using Roche C-peptide CalSet calibration material (Roche Diagnostics, Mannheim, Germany), traceable to WHO International Reference Reagent (IRR) for C-peptide of human insulin for immunoassay (IRR code 84/510) [37]. Quality Control was performed on each day of analysis using low level (0.67 nmol/L) and high level (3.33 nmol/L) PreciControlMultiAnalyte.

3.2.1 Measuring Range

The measuring range of the urine C-peptide assay is from 0.03 to 133 nmol/L.

There is approximately 30 % cross-reactivity with pro-insulin (*see* **Note 3**)

3.3 Roche Assay Used to Measure Creatinine in Urine

Creatinine is analyzed on the Roche P modular analyzer (Roche Diagnostics, Mannheim, Germany). The test principle is a kinetic colorimetric assay based on the Jaffé reaction as described by Popper et al., 1937 and modified by Bartels et al. in 1972 [38].

In alkaline solution, creatinine forms a yellow-orange complex with picrate. The color intensity is directly proportional to the concentration of creatinine in the urine sample and is measured photometrically at 500 nm.

The assay was calibrated on each day of analysis using a two point calibration with Roche Calibrator for Automated Systems calibration material, traceable to Isotope Dilution-Mass Spectrometry (ID-MS) reference method. Quality Control was performed on each day of analysis using Randox assayed urine control level 2 and 3.

3.3.1 Measuring Range, Units

The measuring range for creatinine in urine is 0.36–57.5 mmol/L. Samples that have a creatinine concentration greater than this may be diluted 1:2 via the automatic rerun function, using saline as a diluent.

3.4 Interpretation of Results

UCPCR is restricted to use in patients on insulin treatment to assess endogenous insulin secretion. Its role in patients not on insulin treatment is limited.

If the of UCPCR result is not in keeping with other clinical finding then we would recommend repeating the test especially if it is unexpectedly low. Patients tipping out boric acid preservative from urine collection tube, in a sample taking more than 3 days to reach the laboratory can result in artificially low results [26]. We are not aware of any reason or experience a falsely high value.

Most of the studies have been performed in patients with normal renal function (eGFR > 60 mL/min/1.73 m^2) but it has been validated in patients with Type 2 diabetes with moderate renal impairment (eGFR 30–60 mL/min/1.73 m^2) [29]. The test is unlikely to be appropriate in patients with severe renal impairment.

3.4.1 What Values are Expected in the Different Subtypes?

The UCPCR result is best measured on a post prandial sample taken approximately two hours after a meal stimulus. The interpretation depends on the specific clinical scenario, i.e., type of diabetes and treatment. Median and 5–95th percentile UCPCR values for nondiabetic controls, long standing Type 1 diabetes (>5 years), Type 1 diabetes in the first 5 years after diagnosis, Type 2 diabetes (OHA and insulin treated) and patients with a genetic diagnosis of Maturity Onset Diabetes of the Young (MODY) are shown in Table 1.

Table 1
UCPCR ranges in diabetes subtypes and controls (unpublished data)

Patient group	Males UCPCR (nmol/mmol)					Females UCPCR (nmol/mmol)				
	5th	25th	50th	75th	95th	5th	25th	50th	75th	95th
Controls	0.58	1.64	2.84	7.04	10.39	1.82	3	4.04	6.99	10.37
Type 1 diabetes										
>5 years duration	<0.02	<0.02	0.02	0.02	0.02	0.00	0.00	0.02	0.04	0.04
<5 years duration	0.02	0.55	1.24	1.79	5.78	0.02	0.55	1.24	1.79	5.78
Type 2 diabetes										
On OHA	0.35	1.6	2.87	4.08	7.80	1.28	2.34	3.85	5.68	9.43
On insulin	0.08	0.5	1.3	2.36	5.65	0.15	0.6	1.4	2.8	6.12
HNF 1/4A										
On OHA	0.54	1.36	1.84	2.80	6.10	0.54	1.23	2.93	4.04	10.02
On insulin	0.10	0.54	1.12	1.72	3.47	0.10	0.54	1.12	1.72	3.47

4 Notes

1. **Agreement of urine C-peptide measurements in different commercially available assays**—There is poor comparability between commercial and in-house C-peptide assays making transferability of C-peptides values/cutoffs from research to clinical practice problematic [39–41]. International working groups (in the USA and Europe) are currently addressing the widely disparate values between analytical methods for both C-peptide and insulin [39, 41, 42]. The groups are combining efforts to improve standardization with the aim of establishing a complete reference measurement system and certified primary reference materials based on pure biosynthetic insulin and C-peptide [42].

2. **Using other diluents in the Roche C-peptide assay**—C-peptide is approximately 20-fold more concentrated in urine samples and therefore require dilution before analysis. We have found that the diluent must be proteinaceous (e.g., equine serum albumin) in order to achieve both a concentration and matrix that will generate a comparable and reproducible result. This has been shown for three different platforms measuring C-peptide (Roche, Immulite, and Centaur). (*unpublished data derived at the Blood Sciences Dept., RD&E).

3. **Cross reactivity of the Roche assay with proinsulin—**Proinsulin does cross react with the Roche assay (30 % cross reactivity) despite the fact that the assay utilizes two site monoclonal antibodies. However, proinsulin and split products exist in much lower concentrations than C-peptide in serum (<2 % in nondiabetics) and are therefore of little clinical significance except for in rare conditions such as insulinoma [43]. In addition, the total quantity of C-peptide excreted in the urine per day represents 5–10 % of pancreatic secretion, compared to only 0.05 % of secreted proinsulin (2–3 times lower than insulin of which only 0.1 % is excreted into the urine [44]). With only 30 % cross reactivity in the Roche assay the contribution of proinsulin to urine C-peptide measurement would be negligible.

4. **Making a diagnosis in insulin treated patients—**The principal role of urinary C-peptide is to identify insulin insufficiency, a feature of long term Type 1 diabetes, in insulin treated patients. We have found that >95 % Type 1 diabetes (>5 year duration) have a UCPCR value < 0.2 nmol/mmol.

   - *Type 1 v MODY—*A cutoff UCPCR of 0.2 nmol/mmol differentiates HNF1A/4A MODY from Type 1 diabetes with a sensitivity of 97 % and specificity of 96 % (ROC 0.98) [14].
   - *Type 1 v Type 2—*A cutoff of 0.2 nmol/mmol differentiates Type 2 diabetes from Type 1 diabetes with 94 % sensitivity and 94 % specificity (ROC AUC 0.94).

5. **Monitoring the honeymoon period in Type 1 diabetes—**In patients with Type 1 diabetes it is possible to measure the extent to which they are progressing through the honeymoon phase. UCPCR is highly correlated with the serum C peptide in a formal mixed tolerance test [31].

6. **Assessing in insulin treated Type 2 diabetes the extent of endogenous insulin secretion—**Endogenous insulin secretion can be measured in Type 2 diabetes. UCPCR is highly correlated with the serum C-peptide in a formal mixed tolerance test [28]. Patients with a high endogenous secretion >75th centile are likely be those that benefit most from metformin and other oral agents. Patients with low endogenous insulin secretion <25th centile and particularly if UCPCR <0.02 nmol/mmol will most likely require insulin therapy.

Acknowledgements

Tim McDonald is a National Institute of Health Research CSO funded scientist. The views expressed are those of the author(s) and not necessarily those of the NHS, the NIHR, or the Department of Health.

References

1. Clark PM (1999) Assays for insulin, proinsulin (s) and C-peptide. Ann Clin Biochem 36(Pt 5):541–564
2. Horwitz DL, Rubenstein AH, Katz AI (1977) Quantitation of human pancreatic beta-cell function by immunoassay of C-peptide in urine. Diabetes 26:30–35
3. Matthews DR, Rudenski AS, Burnett MA, Darling P, Turner RC (1985) The half-life of endogenous insulin and C-peptide in man assessed by somatostatin suppression. Clin Endocrinol 23:71–79
4. Duckworth WC, Bennett RG, Hamel FG (1998) Insulin degradation: progress and potential. Endocr Rev 19:608–624
5. Hattersley A, Bruining J, Shield J, Njolstad P, Donaghue K (2006) ISPAD clinical practice consensus guidelines 2006–2007. The diagnosis and management of monogenic diabetes in children. Pediatr Diabetes 7:352–360
6. Koskinen P, Viikari J, Irjala K, Kaihola HL, Seppala P (1986) Plasma and urinary C-peptide in the classification of adult diabetics. Scand J Clin Lab Invest 46:655–663
7. Gjessing HJ, Matzen LE, Faber OK, Froland A (1989) Fasting plasma C-peptide, glucagon stimulated plasma C-peptide, and urinary C-peptide in relation to clinical type of diabetes. Diabetologia 32:305–311
8. Berger B, Stenstrom G, Sundkvist G (2000) Random C-peptide in the classification of diabetes. Scand J Clin Lab Invest 60:687–693
9. Service FJ, Rizza RA, Zimmerman BR, Dyck PJ, O Brien PC, 3rd Melton LJ (1997) The classification of diabetes by clinical and C-peptide criteria. A prospective population-based study. Diabetes Care 20:198–201
10. Palmer JP, Fleming GA, Greenbaum CJ et al (2004) C-peptide is the appropriate outcome measure for type 1 diabetes clinical trials to preserve beta-cell function: report of an ADA workshop, 21-22 October 2001. Diabetes 53:250–264
11. Byrne MM, Sturis J, Fajans SS et al (1995) Altered insulin secretory responses to glucose in subjects with a mutation in the MODY1 gene on chromosome 20. Diabetes 44:699–704
12. Hattersley A, Bruining J, Shield J, Njolstad P, Donaghue KC (2009) The diagnosis and management of monogenic diabetes in children and adolescents. Pediatr Diabetes 10(Suppl 12):33–42
13. Ellard S, Bellanne-Chantelot C, Hattersley AT (2008) Best practice guidelines for the molecular genetic diagnosis of maturity-onset diabetes of the young. Diabetologia 51:546–553
14. Besser RE, Shepherd MH, McDonald TJ et al (2011) Urinary C-peptide creatinine ratio is a practical outpatient tool for identifying hepatocyte nuclear factor 1-{alpha}/hepatocyte nuclear factor 4-{alpha} maturity-onset diabetes of the young from long-duration type 1 diabetes. Diabetes Care 34:286–291
15. Besser RE, Shields BM, Hammersley SE et al (2013) Home urine C-peptide creatinine ratio (UCPCR) testing can identify type 2 and MODY in pediatric diabetes. Pediatr Diabetes 14:181–188
16. Bolner A, Lomeo L, Lomeo AM (2005) "Method-specific" stability of serum C-peptide in a multicenter clinical study. Clin Lab 51:153–155
17. Greenbaum CJ, Mandrup-Poulsen T, McGee PF et al (2008) Mixed-meal tolerance test versus glucagon stimulation test for the assessment of beta-cell function in therapeutic trials in type 1 diabetes. Diabetes Care 31:1966–1971
18. Assayfinder. Accessed 28 Jul 2011 from www.assayfinder.com
19. McDonald TJ, Perry MH, Peake RW et al (2012) EDTA improves stability of whole blood C-peptide and insulin to over 24 hours at room temperature. PLoS One 7:e42084
20. Gjessing HJ, Matzen LE, Froland A, Faber OK (1987) Correlations between fasting plasma C-peptide, glucagon-stimulated plasma C-peptide, and urinary C-peptide in insulin-treated diabetics. Diabetes Care 10:487–490
21. Huttunen NP, Knip M, Kaar ML, Puukka R, Akerblom HK (1989) Clinical significance of urinary C-peptide excretion in children with insulin-dependent diabetes mellitus. Acta Paediatr Scand 78:271–277
22. Meistas MT, Zadik Z, Margolis S, Kowarski AA (1981) Correlation of urinary excretion of C-peptide with the integrated concentration and secretion rate of insulin. Diabetes 30:639–643
23. Aoki Y (1991) Variation of endogenous insulin secretion in association with treatment status: assessment by serum C-peptide and modified urinary C-peptide. Diabetes Res Clin Pract 14:165–173
24. Cote AM, Firoz T, Mattman A, Lam EM, von Dadelszen P, Magee LA (2008) The 24-hour urine collection: gold standard or historical practice? Am J Obstet Gynecol 199(625):e621–e626
25. Hoogwerf BJ, Barbosa JJ, Bantle JP, Laine D, Goetz FC (1983) Urinary C-peptide as a

measure of beta-cell function after a mixed meal in healthy subjects: comparison of four-hour urine C-peptide with serum insulin and plasma C-peptide. Diabetes Care 6:488–492

26. McDonald TJ, Knight BA, Shields BM, Bowman P, Salzmann MB, Hattersley AT (2009) Stability and reproducibility of a single-sample urinary C-peptide/creatinine ratio and its correlation with 24-h urinary C-peptide. Clin Chem 55:2035–2039

27. Oram RA, Rawlingson A, Shields BM et al (2013) Urine C-peptide creatinine ratio can be used to assess insulin resistance and insulin production in people without diabetes: an observational study. BMJ Open 3:e003193

28. Jones AG, Besser RE, McDonald TJ et al (2011) Urine C-peptide creatinine ratio is an alternative to stimulated serum C-peptide measurement in late-onset, insulin-treated diabetes. Diabet Med 28:1034–1038

29. Bowman P, McDonald TJ, Shields BM, Knight BA, Hattersley AT (2012) Validation of a single-sample urinary C-peptide creatinine ratio as a reproducible alternative to serum C-peptide in patients with Type 2 diabetes. Diabet Med 29:90–93

30. Besser RE (2013) Determination of C-peptide in children: when is it useful? Pediatr Endocrinol Rev 10:494–502

31. Besser RE, Ludvigsson J, Jones AG et al (2011) Urine C-peptide creatinine ratio is a noninvasive alternative to the mixed-meal tolerance test in children and adults with type 1 diabetes. Diabetes Care 34:607–609

32. Besser RE, Shields BM, Casas R, Hattersley AT, Ludvigsson J (2013) Lessons from the mixed-meal tolerance test: use of 90-minute and fasting C-peptide in pediatric diabetes. Diabetes Care 36:195–201

33. Hope SV, Jones AG, Goodchild E et al (2013) Urinary C-peptide creatinine ratio detects absolute insulin deficiency in Type 2 diabetes. Diabet Med 30:1342–1348

34. Thomas NJ, Shields BM, Besser RE et al (2012) The impact of gender on urine C-peptide creatinine ratio interpretation. Ann Clin Biochem 49:363–368

35. Ashby JP, Frier BM (1981) Circulating C peptide: measurement and clinical application. Ann Clin Biochem 18:125–130

36. Koskinen P (1988) Nontransferability of C-peptide measurements with various commercial radioimmunoassay reagents. Clin Chem 34:1575–1578

37. Bristow AF, Das RE (1988) WHO international reference reagents for human proinsulin and human insulin C-peptide. J Biol Stand 16:179–186

38. Bartels H, Bohmer M, Heierli C (1972) Serum creatinine determination without protein precipitation. Clin Chim Acta 37:193–197

39. Wiedmeyer HM, Polonsky KS, Myers GL et al (2007) International comparison of C-peptide measurements. Clin Chem 53:784–787

40. Little RR, Rohlfing CL, Tennill AL et al (2008) Standardization of C-peptide measurements. Clin Chem 54:1023–1026

41. Marcovina S, Bowsher RR, Miller WG et al (2007) Standardization of insulin immunoassays: report of the American Diabetes Association Workgroup. Clin Chem 53:711–716

42. Miller WG, Thienpont LM, Van Uytfanghe K et al (2009) Toward standardization of insulin immunoassays. Clin Chem 55:1011–1018

43. Vauhkonen IK, Niskanen LK, Mykkanen L, Haffner SM, Uusitupa MI, Laakso M (2000) Hyperproinsulinemia is not a characteristic feature in the offspring of patients with different phenotypes of type II diabetes. Eur J Endocrinol 143:251–260

44. Constan L, Mako M, Juhn D, Rubenstein AH (1975) The excretion of proinsulin and insulin in urine. Diabetologia 11:119–123

Part III

Immunopathogenesis

Methods in Molecular Biology (2016) 1433: 105–117
DOI 10.1007/7651_2015_287
© Springer Science+Business Media New York 2015
Published online: 23 January 2016

Histology of Type 1 Diabetes Pancreas

Abby Willcox and Kathleen M. Gillespie

Abstract

The islets of Langerhans play a critical role in glucose homeostasis. Islets are predominantly composed of insulin-secreting beta cells and glucagon-secreting alpha cells. In type 1 diabetes, the beta cells are destroyed by autoimmune destruction of insulin producing beta cells resulting in hyperglycemia. This is a gradual process, taking from several months to decades. Much of the beta cell destruction takes place during a silent, asymptomatic phase. Type 1 diabetes becomes clinically evident upon destruction of approximately 70–80 % of beta cell mass. Studying the decline in beta cell mass and the cells which are responsible for their demise is difficult as pancreatic biopsies are not feasible in patients with type 1 diabetes. The relative size of islets and their dispersed location throughout the pancreas means in vivo imaging of human islets is currently not manageable. At present, there are no validated biomarkers which accurately track the decline in beta cell mass in individuals who are at risk of developing, or have already developed, type 1 diabetes. Therefore, studies of pancreatic tissue retrieved at autopsy from donors with type 1 diabetes, or donors with high risk markers of type 1 diabetes such as circulating islet-associated autoantibodies, is currently the best method for studying beta cells and the associated inflammatory milieu in situ. In recent years, concerted efforts have been made to source such tissues for histological studies, enabling great insights to be made into the relationship between islets and the inflammatory insult to which they are subjected. This article describes in detail, a robust immunohistochemical method which can be utilized to study both recent, and archival human pancreatic tissue, in order to examine islet endocrine cells and the surrounding immune cells.

Keywords: Type 1 diabetes, Beta cell, Alpha cell, Islet, Pancreas, Histology, Immunohistochemistry, Insulitis, DAB, Antibody

1 Introduction

The human pancreas is a multifunctional organ. The exocrine compartment comprises acinar cells which produce and secrete digestive enzymes into the pancreatic ductal system which are then drained into the gut. The endocrine compartment synthesizes and secretes glucose homeostasis hormones which are secreted into the circulation in response to blood glucose. The exocrine compartment comprises approximately 98 % of pancreatic mass, with the endocrine comprising just 2 %. The endocrine cells of the pancreas are clustered into discrete "islands" known as islets of Langerhan, which are scattered throughout the pancreas. Islets are composed of several different endocrine cell types; alpha cells which secrete glucagon, beta cells which secrete insulin, delta cells which secrete somatostatin, PP cells which secrete pancreatic polypeptide, and a minor fraction of epsilon cells which secrete ghrelin. Islets are highly vascularized by capillaries and are also innervated by the sympathetic, parasympathetic, and sensory nervous system [1]. Islets are approximately 150–300 μm in diameter and contain approximately 1500 cells each [2].

The autoimmune destruction of beta cells results in the development of type 1 diabetes. The inflammatory infiltrate which targets the islets is known as "insulitis." The wide distribution of islets throughout the organ makes the study of the endocrine compartment of the pancreas very difficult as the islets are too small to resolve by in vivo imaging in humans. Advances are being made in the field of in vivo imaging of pancreatic beta cells and insulitis [3] but information on beta cell function can currently only be gained indirectly by measurement of islet hormones in the circulation, or by measuring the 31-amino-acid C-peptide, produced in equimolar quantities when proinsulin is cleaved into insulin, in the circulation or the urine. Blood glucose levels following a meal can also be measured in order to gain information on pancreatic function but this does not correlate with beta cell mass as glucose homeostasis can be maintained with as little as 30 % of normal beta cell mass [4, 5]. These issues are further complicated by the heterogeneous preclinical phase of the disease during which beta cell destruction can take place over months up to several years and is unique to each individual. Circulating islet associated autoantibodies are useful biomarkers which indicate that beta cell autoimmunity is taking place and aid in determining risk of developing type 1 diabetes [6]. They offer no insight however into the extent of beta cell destruction or the cells which are responsible for it.

Study of the human pancreas using histological methods is therefore extremely important as it provides direct insights into the autoimmune process in the type 1 diabetic pancreas. Such tissue is generally only available at autopsy as taking pancreatic biopsies is a complicated procedure and there are many issues, both ethical and practical, with taking pancreatic samples from individuals with type 1 diabetes [7]. Immunohistochemical staining of human pancreas tissue sections for various endocrine, immune cell and proliferative markers provides valuable information on the state of the islets in terms of beta cell number remaining at various stages of disease, which cells are involved in beta cell death, and importantly, whether there are any indications that beta cell mass may be able to recover or regenerate following the autoimmune attack.

This chapter aims to describe a robust immunohistochemistry method which can be applied to new and archival pancreatic tissues in order to reveal important features of the pancreatic islets which can be examined and quantified in order to gauge the degree of beta cell loss and the severity of any ongoing autoimmune insult in the pancreas of donors with new onset type 1 diabetes.

Pancreatic tissue from individuals with type 1 diabetes, in particular new onset type 1 diabetes, is an extremely rare resource. Availability of tissue sections is limited and therefore, it is very important that all staining procedures are optimized on control tissues which are available in greater abundance. Nondiabetic pancreatic tissue is recommended for islet endocrine cell antibodies, and human spleen or tonsil tissue for antibodies raised against immune cell or proliferation markers. This article assumes the use of formalin-fixed, paraffin-embedded (FFPE) tissue sections which have been thinly sectioned (4–8 μm) with a microtome onto standard glass microscope slides, however, the adaptation of the method for the use of snap-frozen sections will be discussed in Sect. 4. Suggestions are also made as to which antibodies can be used for a general examination of the type 1 diabetic pancreas, however any primary antibody of interest can be switched into this protocol and therefore, the selection of secondary antibodies outside of the scope of those described below, and all relevant control reagents, should be identified and sourced by the end user.

2 Materials

1. Xylene.

2. Absolute ethanol (use distilled water (dH$_2$O) to make dilutions of 90 %, 70 %, and 50 % ethanol).

3. Antigen retrieval buffers: Tris-EDTA pH 9.0: 10 mM Tris-base, 1 mM EDTA in dH$_2$O, or, citrate buffer pH 6.0: 100 mM citric acid in dH$_2$O.

4. Tris-buffered saline (TBS) pH 7.4: 50 mM Tris, 150 mM NaCl in dH$_2$O.

5. A microwave suitable for laboratory use.

6. A microwavable pot with a vented lid, large enough to hold 1 L of antigen retrieval buffer and completely immerse a rack of microscope slides (standard slide dimensions 26 mm × 76 mm × 1 mm).

7. Glass Coplin jars (standard 10-slide Coplin jars hold approximately 50 ml of solution).

8. An incubation chamber with a lid in which glass microscope slides can be laid flat and slightly raised from the bottom of the chamber. Place a dampened paper towel along one edge of the chamber in order to maintain humidity and prevent drying out of the tissue sections. A large, shallow Tupperware pot with 5 ml serological pipettes firmly attached to the bottom in rows will suffice.

9. Blocking reagent: TBS supplemented with 5 % normal serum (Vector Laboratories).

10. Endogenous peroxidase 3 % blocking solution: H_2O_2 stock reagent (30 %) diluted 1/10 with dH_2O.

11. Primary antibodies to detect islet beta and alpha cells respectively: guinea-pig anti-insulin (Dako; A0564), rabbit anti-glucagon (Abcam; ab18461). Dilute in either TBS + 0.5 % normal serum or commercial diluents can be purchased, e.g., Dako REAL™ Antibody Diluent.

12. Primary antibodies to detect immune cell infiltrates, including CD45 (pan-lymphocyte), CD3 (pan-T cell), CD4 (helper T cells), CD8 (cytotoxic T cells), CD20 (B cells), and CD68 (macrophages), have been described in detail previously [8].

13. Isotype control antibodies: immunoglobulin (Ig) of the same isoform and species of the primary antibody, e.g., mouse IgG1 or, IgG2a (available from multiple sources).

14. Dako Envision™ Detection Systems Peroxidase/DAB, Rabbit/Mouse secondary detection kit: contains a ready-to-use cocktail of goat anti-rabbit IgG and goat anti-mouse IgG polymer-horseradish peroxidase (HRP) conjugated secondary antibodies, and DAB+ chromogen and DAB+ substrate buffer. The secondary antibody cocktail is also able to detect guinea-pig IgG so can be used as a universal secondary antibody for all three species of primary antibody.

15. Mayer's hematoxylin. Filter the hematoxylin through Whatman filter paper before use.

16. Blueing solution; Scots Tap Water Substitute (STWS): 3.5 g sodium bicarbonate, 20 g magnesium sulfate, 1 L dH_2O.

17. DPX mounting medium.

18. Glass coverslips, 25 × 60 mm.

3   Method

3.1   Setting Up Antibody Controls

As with any method which involves antibody labeling, appropriate controls must be employed to confirm that both primary and secondary antibodies are binding specifically. These are very important when staining human pancreatic tissue as islets can be "sticky" and are prone to staining with a wash of light nonspecific background when using immuno-peroxidase techniques. This needs to be distinguishable from true immunopositive signal.

To control for primary antibody specificity, an antibody pre-quenching step using the relevant immunizing peptide or protein, should be employed. Immunizing peptides and proteins can be purchased from the manufacturer of which the antibodies were

Primary antibody quality control 1:
Blocking peptide/protein

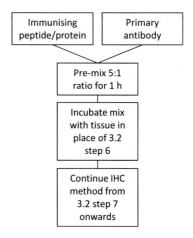

Fig. 1 The flowchart demonstrates how to adapt the method to include a control condition under which the primary antibody would be quenched by immunizing peptide or protein prior to incubation with the tissue. If positive labeling persists under these conditions, it suggests that the primary antibody exhibits off-target binding properties

sourced. Premix the primary antibody with its corresponding peptide/protein in TBS at a ratio of at least 1:5 for 1 h with frequent agitation, followed by a brief centrifugation. Apply the premixed solution to the section in the same way that primary antibody alone would be incubated (*see* Sect. 3.2, **step 6**) (Fig. 1). If the primary antibody is specific for the immunizing protein, it should be quenched during the premixing step and unable to bind to the tissue section. This should result in the absence of positive staining. A second quality control which can be employed in certain circumstances in human tissue is the use of alternative tissues which are known not to express the protein of interest. This is easily accomplished for insulin staining as the protein is only expressed in the pancreas. However, this control step may not be feasible if the protein of interest is widely expressed in many cell types, or if its expression profile is ambiguous.

There are two quality control steps which should be employed to confirm the specificity of secondary antibodies. The first requires replacing the primary antibodies with isotype control antibodies. These are irrelevant Ig of the same isotype and derived from the same host species of the primary antibody, e.g., mouse IgG1 or, IgG2a (Fig. 2a). Replacing the primary antibody with isotype control Ig should result in a lack of positive staining as the control Ig should target any epitopes in the human tissue and therefore, should not bind. In this case, the secondary antibody should have no primary antibody to bind to. If, under these control conditions

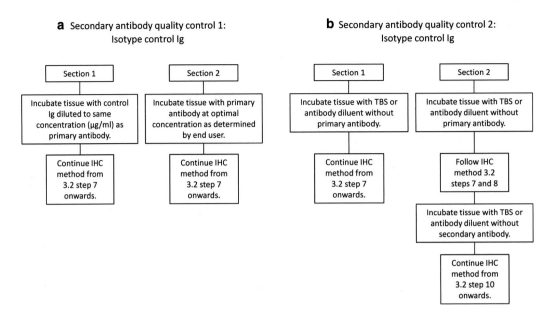

Fig. 2 (**a**) The flowchart demonstrates how to include the quality control condition in which primary antibody is replaced by Ig of the same isotype and host species. If positive labeling persists under these conditions, it suggests that the secondary antibody exhibits off-target binding properties, or, in some cases the Ig itself may have bound to the tissue. (**b**) demonstrates an additional control measure which should always be employed. It will also determine the cause of any nonspecific staining which presents under the control conditions described in Figs. 1a and 2a

there is still staining then it suggests that the secondary antibody is binding nonspecifically to the tissue. The second quality control step for secondary antibody specificity requires (a) the incubation of one tissue section with only TBS or antibody diluent without addition of primary antibody (Sect. 3.2, **step 6**) and (b) an adjacent tissue section incubated with only TBS or antibody diluent in place of both the primary and secondary antibodies (Sect. 3.2, **steps 6 and 9**, respectively) (Fig. 2b). Should positive staining develop under condition "a" but not condition "b" then it points to a nonspecificity issue with the secondary antibody. However, if positive staining develops under conditions "a" and "b," it suggests that endogenous peroxidase activity in the tissue was not completely quenched by the endogenous peroxidase 3 % blocking solution and has reacted with the DAB reagent. To overcome this issue a higher concentration of H_2O_2 or a slightly increased incubation time will be required (e.g., increase from 5 min to 10 min).

3.2 Immuno-histochemistry Staining Method

Steps **1–2** and **17–20** must be carried out in a fume cupboard.

1. Place the slides in a Coplin jar filled with Xylene to fully immerse the tissue sections, for 5 min. Repeat with a second Coplin jar of Xylene. This step will dissolve the wax and remove it from the tissue.

2. Bring the slides through a series of Coplin jars containing decreasing concentrations of ethanol (100 %, 90 %, 70 % and 50 %), and finally into a Coplin jar of dH_2O. This step will rehydrate the tissue, enabling the binding of antibodies in subsequent steps.

3. Gaining access to antigen binding epitopes within FFPE tissue can be problematic. Cross-links formed during formalin fixation can physically block antibodies from accessing their cognate epitope. The process of antigen retrieval, also known as heat-mediated epitope retrieval (HIER), is often required to break down the cross-links. If this method is modified for the use of fresh-frozen tissues then HIER (**step 3a–d**) should be omitted (*see* **Note 1**).

   (a) Fill a microwaveable pot with 1 L of antigen retrieval buffer. The choice of antigen retrieval buffer is specific to each primary antibody, and must be determined by the end user following optimization experiments. There is not one buffer which optimally unmasks epitopes for all antibodies (*see* **Note 2**).

   (b) Place the slides into a microwavable plastic slide rack and immerse fully in the antigen retrieval buffer.

   (c) Secure the lid onto the pot, while allowing steam to escape through a vent or small hole in the lid.

   (d) Place the pot containing the slides into a microwave and set to full power ("high" on an 800 W microwave) for 20 min, then remove the pot from the microwave, loosen the lid and leave the sections in the buffer to cool for 20 min.

4. Remove the slides from the antigen retrieval buffer one at a time, being careful not to allow the tissue sections to dry out as the slides will still be warm. Lay the slide flat, tissue side up, in the incubation chamber and immerse the tissue section in 100–500 µl (depending on the size of the section, use the same volume for all incubation chamber steps) of blocking reagent, ensuring it is fully covered with solution. When choosing serum for the blocking reagent it is optimal to use normal serum from the host species in which the secondary antibodies are raised, or a serum which is irrelevant to any of your primary antibody species (e.g., if you are using a primary antibody which was raised in rabbit then you should never use normal rabbit serum for blocking as this will generate a blanket of nonspecific staining). Place the lid on the chamber and incubate for 30 min at room temperature.

5. Pour the blocking reagent off of the sections into the bottom of the tray and lay the slides down flat.

6. Incubate the sections with primary antibodies diluted appropriately (as suggested by the antibody data sheet combined with end user optimization experiments) in antibody diluent, in the same manner as described for the blocking reagent. Sections can be incubated for 1–2 h at room temperature or, overnight at 4 °C, depending on the individual requirements of each primary antibody (e.g., if the antigen is low in abundance, the tissue autolysed, or the primary antibody only available at low concentration, then a longer incubation time is often required (*see* **Note 3**)).

7. Pour the primary antibody solution off of the sections and place the slides in a Coplin jar filled with TBS for 5 min. Repeat twice more to wash all unbound primary antibody from the sections.

8. Lay the slides flat in the incubation chamber and incubate with endogenous peroxide blocking solution for 5 min. Pour off the solution and wash once more in a Coplin jar of TBS for 5 min to stop the reaction. If this method is being employed for immunofluorescence staining, this step should be omitted and the modified method described in **Note 4** should be followed from this stage onwards.

9. Lay the slides back in the incubation chamber and immerse with secondary antibody for 1 h. The Envision™ secondary antibody described in **step 13** of Sect. 2 is ready-to-use. Approximately 3–5 drops will be suitable to immerse most tissue sections. If using a concentrated stock HRP-conjugated secondary antibody, ensure that it is targets the relevant primary antibody host species and dilute in the same manner as described for primary antibodies (Sect. 2, **step 11**).

10. Pour the secondary antibody solution off of the sections and place the slides in a Coplin jar of TBS for 5 min. Repeat twice more to wash all unbound antibody from the sections.

11. Using the Envision™ Detection Systems Peroxidase/DAB kit (described in Sect. 2, **step 14**), dilute the DAB+ chromogen in the DAB+ substrate buffer at a ratio of 1:50 and mix well.

12. To develop the chromogen, enabling visualization of the location of the protein of interest, lay each of the slide flat in the incubation chamber and incubate with the mixed DAB+ solution for 10 min (*see* **Note 5**). When labeling antigens which are expressed in abundance, it is possible to visualize a brown precipitate forming by eye as the antibody-bound peroxidase oxidases the 3,3′-diaminobenzidine. If staining for insulin or glucagon on nondiabetic pancreas sections, brown "spots" should appear throughout the section, highlighting the location of the islets (Fig. 3a, b). Antigens which are expressed in low abundance may not be visible without the aid of a microscope. If the entire section turns a shade of light–dark brown it

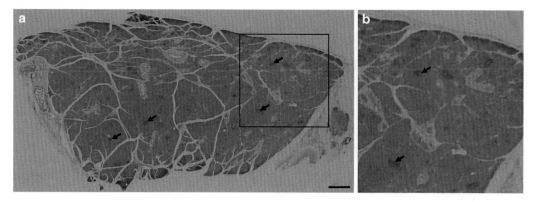

Fig. 3 (**a**) A section of human pancreas stained with rabbit anti-glucagon primary antibody and counterstained with hematoxylin (*blue areas*) using the method described in this chapter. Focal areas of brown staining which represent glucagon-positive alpha cells, and therefore islets, are visible by eye. A selection of such islets is highlighted by the *black arrows*. The region highlighted by the *black box* is shown at higher magnification in image (**b**). Scale bar, 1000 μm

suggests that too high a concentration of primary antibody was used or that the blocking steps were not completely effective.

13. Following the DAB incubation, dispose of the DAB solution as per the material safety data sheet suggests. Place the slides in a Coplin jar of dH$_2$O for 5 min to inhibit any further oxidization reaction.

14. To counterstain the tissue a purple/blue color, place the slides in a Coplin jar of filtered Mayer's hematoxylin for 1–2 min.

15. Place the slides into a Coplin jar of tap water and leave under a gently running tap for 5 min. The color of the sections will change when immersed in tap water to a slightly darker shade of blue/purple due to the slightly acidic pH of tap water.

16. The counterstained sections can optionally be placed in a Coplin jar of STWS blueing solution for 10 s. This step will induce further blueing of the sections and increase the sharpness of the counterstain.

17. To complete the procedure, dehydrate the sections by placing the slides in 50 %, 70 %, 90 % and 2 × 100 % ethanol for 5 min each. It is crucial that all water in the tissue is displaced by ethanol to enable effective mounting and preservation of the sections.

18. Place the slides in xylene for 5 min, and repeat. The xylene will expel the ethanol from the tissue, and is also suitable for mixing with the DPX mounting medium, while ethanol is not.

19. Lay the slides flat on a paper towel. Using a plastic Pasteur pipette, apply approximately 600 μl of DPX mounting medium to the section. Gently lower a glass coverslip at a diagonal angle over the slide, using slow movements to prevent air bubbles

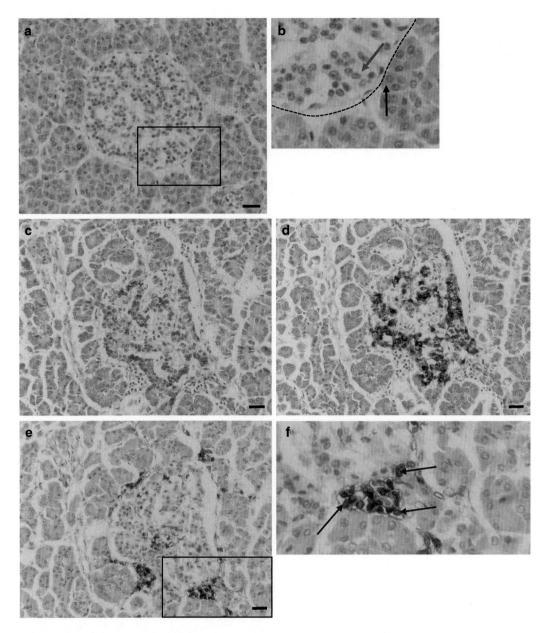

Fig. 4 Images of human pancreatic islets in situ. (**a**) shows an islet stained with hematoxylin (*blue*), surrounded by acinar cells. (**b**) highlights the area outlined by the *black box* in (**a**). It shows the distinction between the lightly stained cytoplasm of the islet cells (*orange arrow*) and the darker stained cytoplasm of the acinar cells (*black arrow*). The boundary between the islet and exocrine compartments is marked by the *dashed black line*. Images (**c**), (**d**) and (**e**) show islets stained for glucagon, insulin, and the pan-lymphocyte marker CD45, respectively. Image (**f**) shows a highlighted area represented by the *black box* in image (**e**). CD45-immunopositive lymphocytes are visible as dark nuclei with minimal visible cytoplasm. Scale bars, 20 μm

forming and being trapped on the section. Turn the slide over and apply gentle pressure to spread out the mounting medium evenly and remove any excess. Turn the slide back upright and square up the coverslip with the slide.

20. Lay the slides down flat to dry overnight.

3.3 Imaging

The tissue sections can be visualized using any microscope with bright-field capability. If image capture is required then a bright-field microscope with a mounted color camera linked to appropriate image capture software is necessary.

Islets can be identified histologically as discrete structures which are embedded within the exocrine tissue and are surrounded by a basement membrane [9]. As described above, islets stain a lighter shade with hematoxylin than the surrounding acinar cells (Fig. 4a, b). While the whole islet can be readily identified based on hematoxylin staining alone, in human pancreas immunohistochemical labeling of the islet hormones is crucial in order to confidently identify individual cell types as this cannot be done based on location alone. In humans, the alpha and beta cells do not display clear differential localization as seen in mouse islets (i.e., an alpha cell mantle with a beta cell core) [10]. Therefore, to fully establish the beta cell content of the islets or the alpha cell–beta cell ratio, cell-type specific staining must be employed (Fig. 4c, d). Insulitis can be identified histologically as small mononuclear cells, the nuclei of which stain intensely with hematoxylin, with very little or no visible cytoplasm (Fig. 4e, f). These cells represent the lymphocyte population, however there is no way of distinguishing between lymphocytes sub-types histologically so subtype specific antibodies must be employed to reveal the composition of the infiltrate. This is also important for detection of the non-lymphocyte component of the infiltrate which is not so readily identifiable in the pancreas using histology alone.

4 Notes

1. The described method can be easily modified for either fixed-frozen or fresh-frozen tissue sections. For fixed sections, wash the sections in TBS 3×5 min to remove any OCT cutting medium and then proceed from **step 4** of the method (incubating with blocking reagent). If using fresh frozen sections, bring them to room temperature and dry any condensation from the slides using tissue paper. Lay the slides down flat and incubate with 1 ml of 10 % formalin per slide for 15 min (in a fume cupboard). Wash the sections three times 5 min with TBS and then proceed from **step 4**.

2. Antigen retrieval is a step which often requires careful optimization before application to your test samples. There are many variations on the buffers and method described in this article. However, the two buffers described have consistently produced robust results and are very commonly employed for this purpose. The exact mechanisms which underlie antigen retrieval are not fully understood, however pH has been shown to be very important and extremes of pH are more effective at un-masking epitopes [11]. Each epitope is optimally un-masked under different conditions, therefore conditions for each primary antibody need to be ascertained as they are unlikely to be the same for each primary antibody being used. There are also a wide range of methods used. This article describes the use of a microwave to heat the buffer and sections, however, the use of a pressure cooker has also been found to be very effective. It is also possible to unmask epitopes enzymatically using commercially available kits. Antigen retrieval is not required for frozen sections as they are not subject to the same intense cross-linking which affects FFPE tissue. Also, some antibodies, including the Dako anti-insulin and anti-glucagon antibodies described above, do not require antigen retrieval, therefore this step can be omitted as it is always preferable to expose your tissue sections to as few processing steps as possible to better preserve their integrity.

3. Human pancreatic tissue retrieved at autopsy can be particularly difficult to work with. This is in part due to the digestive nature of the pancreas which makes it susceptible to degradation by autolysis. Given that it is not possible to retrieve the organ immediately following death, even short ischemia times of a few hours can result in some autolysis. Very pronounced autolysis can lead to reduced labeling of some antigens, although fortunately the islets and also the immune infiltrate are reasonably resistant to autolysis in the short term, likely due to protection provided by the double basement membrane and the fact that the pancreatic ducts do not pass through the islets. Labeling issues caused by mild autolysis can often be resolved by an increased incubation time with primary antibody from 1 h to overnight at 4 °C. However, there is no way to improve the disturbed morphology of any affected cells.

4. The method can also easily be adapted for immunofluorescence staining. Firstly, omit the endogenous peroxidase blocking step. Secondly, replace the HRP-conjugated secondary antibody with a relevant fluorescent probe-conjugated secondary antibody, and finally terminate the procedure at **step 10** after washing off the secondary antibody. Mount the sections using a mounting medium appropriate for immunofluorescent staining, such as VECTASHIELD® (Vector Laboratories) or

ProLong® Gold antifade reagent (Life Technologies) and keep them protected from light. If required, mounting medium can be purchased premixed with a nuclear dye such as DAPI or Hoechst to stain all of the cell nuclei. Both of these stains emit fluorescence in the blue spectrum.

5. While 10 min incubation with Dako DAB reagent is the recommended time for optimal development of most reactions, this time can be modified to suit individual end users. Some reactions develop precipitate rapidly and may only require a 2–3 min incubation time, whereas others may take longer. Too long an incubation in DAB reagent will eventually result in nonspecific background staining of the tissue which might mask any specific immunopositive staining. It is important to determine equilibrium between developing a good visual precipitate and minimizing the degree of background staining. The most important aspect of the DAB incubation time is to keep it consistent between experiments. For example, if an incubation time of 5 min is determined for one sample, then this time point must be used consistently for all other samples so as not to gain false information regarding different staining intensity between specimens.

Acknowledgement

This work was supported by a Diabetes UK grant to K.M.G. and a European Foundation for the Study of Diabetes (EFSD) Fellowship to A.W.

References

1. Rodriguez-Diaz R et al (2011) Innervation patterns of autonomic axons in the human endocrine pancreas. Cell Metab 14(1):45–54

2. Pisania A et al (2010) Quantitative analysis of cell composition and purity of human pancreatic islet preparations. Lab Invest 90(11):1661–1675

3. Di Gialleonardo V et al (2012) Imaging of beta-cell mass and insulitis in insulin-dependent (Type 1) diabetes mellitus. Endocr Rev 33(6):892–919

4. Kloppel G et al (1985) Islet pathology and the pathogenesis of type 1 and type 2 diabetes mellitus revisited. Surv Synth Pathol Res 4 (2):110–125

5. Eizirik DL et al (2008) Use of a systems biology approach to understand pancreatic beta-cell death in Type 1 diabetes. Biochem Soc Trans 36(Pt 3):321–327

6. Bingley PJ et al (1997) Prediction of IDDM in the general population: strategies based on combinations of autoantibody markers. Diabetes 46(11):1701–1710

7. Krogvold L et al (2014) Pancreatic biopsy by minimal tail resection in live adult patients at the onset of type 1 diabetes: experiences from the DiViD study. Diabetologia 57 (4):841–843

8. Willcox A et al (2009) Analysis of islet inflammation in human type 1 diabetes. Clin Exp Immunol 155(2):173–181

9. Korpos E et al (2013) The peri-islet basement membrane, a barrier to infiltrating leukocytes in type 1 diabetes in mouse and human. Diabetes 62(2):531–542

10. Bosco D et al (2010) Unique arrangement of alpha- and beta-cells in human islets of Langerhans. Diabetes 59(5):1202–1210

11. Shi SR, Key ME, Kalra KL (1991) Antigen retrieval in formalin-fixed, paraffin-embedded tissues: an enhancement method for immunohistochemical staining based on microwave oven heating of tissue sections. J Histochem Cytochem 39(6):741–748

Methods in Molecular Biology (2016) 1433: 119–125
DOI 10.1007/7651_2015_295
© Springer Science+Business Media New York 2015
Published online: 13 December 2015

Identification of Islet Antigen-Specific CD8 T Cells Using MHCI-Peptide Tetramer Reagents in the Non Obese Diabetic (NOD) Mouse Model of Type 1 Diabetes

James A. Pearson and F. Susan Wong

Abstract

MHCI-peptide tetramer staining is an important technique in order to identify antigen-specific T cells within a heterogeneous cell population. The reagents may be used to isolate antigen-specific T cells and can help identify their role in disease. Here, we describe how to make tetramer from peptide:MHC monomers together with a protocol for staining antigen-specific cell populations with advice on generating a complementary antibody phenotyping panel.

Keywords: MHCI-peptide tetramer staining, Antigen-specific, T cell receptor, Flow cytometry

1 Introduction

Identification of antigen-specific T cell populations by MHCI-peptide tetramer staining has revolutionized studies in infection, cancer, and autoimmunity [1]. Antigen specificity is dictated by the MHC class I folded with the chosen peptide, although there may be some promiscuity with some T cell receptors (TCR) recognizing multiple peptides [2]. There are many existing protocols available for generation of MHCI-peptide-tetramer complexes. Here, we describe how biotinylated peptide:MHC monomers are used to bind to a streptavidin conjugated to a fluorochrome, thus forming a tetramer. This enables detection of low frequency antigen-specific cells in the context of autoimmunity, using flow cytometry. Cell populations containing antigen-specific CD8 T cells are preincubated with Dasatinib (a tyrosine kinase inhibitor), preventing T cell receptor (TCR) recycling from the cell surface which subsequently provides better tetramer staining [3]. Post-tetramer staining, an antibody panel is used to further phenotype the cells. The technique to be discussed has been optimized for detection of low frequency antigen-specific CD8 T cells and includes staining for murine $CD8^+$ T cells specific for insulin B_{15-23} [4] in the Non Obese Diabetic (NOD) mouse model of type 1 diabetes. However this technique is also applicable to other antigen-specific $CD8^+$ T cells in the NOD mouse such as those specific for Islet-specific glucose-6-phosphatase catalytic-subunit related protein amino acids 206-214 (IGRP206–214) [5].

2 Materials

2.1 Tetramer Synthesis Components

- Sterile $1\times$ PBS without calcium and magnesium
- 70 % Ethanol
- Peptide:MHC class I monomers
- Protease Inhibitor Cocktail Set I (Merck Millipore 539131) made to $100\times$ using sterile water.
- High quality Streptavidin bound to a fluorochrome (*see* **Note 1**)— here Streptavidin BV421 (BioLegend 405225)
- 1.5 ml microfuge tubes

2.2 Tetramer Staining Components

- Tetramer
- Freshly isolated cells
- Tetramer Wash Buffer (TWB): 2 % FCS in $1\times$ PBS and store at 4 °C
- FACS tubes
- 10 mg Dasatinib (Axon Medchem, 1392) diluted in DMSO to 1 mM, aliquot into 5 µl aliquots and store at −20 °C until required
- Fcx block (Trustain)
- Antibody Panel (*see* **Note 2**) including viability dye eFluor 780 (eBioscience 65-0865-14), CD4 PeCy7 (eBioscience 25-0042-52), CD19 PerCpCy5.5 (eBioscience 65-0865-14), CD8 FITC (BD 553030), CD11b APC (BD 553312).

2.3 T Cell Receptor Transgenic CD8 T Cells

- CD8 T cells from the $G9C\alpha^{-/-}$ TCR transgenic mice [6] that recognize H-$2K^d$-insulin B_{15-23} peptide

3 Methods

3.1 Tetramer Synthesis Protocol

1. Clean bench and anything that will be handled such as pipettes and styrofoam boxes with 70 % ethanol.
2. Thaw the monomer on ice in a box with a lid on as this will prevent monomer degradation prior to making tetramer and avoid refreezing.

3. A suitable amount should be sufficient for 50 tests (i.e., 50 μg monomer) assuming 1 μg of tetramer will be the optimal amount for staining the sample. When making tetramer, it is important to add the monomer and streptavidin-fluorochrome at a ratio of 4:1 (i.e., for MHC class I monomers 50 μg is approximately equivalent to 1 nmol and therefore a total of 0.25 nmol of streptavidin-fluorochrome will be added). To calculate the amount of monomer required, divide the total amount (in μg) by the concentration of the monomer in μg/μl, i.e., 50 μg/X μg/μl = Volume in μl and add this amount to a sterile microfuge tube on ice.

4. To this aliquot of monomer on ice, add 1 μl of the diluted protease inhibitor and mix well.

5. To calculate the amount of streptavidin-fluorochrome conjugate required, multiply the weight in KDa (for both the streptavidin and fluorochrome) by the nmol required (0.25 nmol), i.e., (52.8 kDa (streptavidin) + X kDa (fluorochrome)) × 0.25 nmol = Volume in μl. This assumes a streptavidin-fluorochrome concentration of 1 mg/ml (*see* **Note 3**).

6. Add 1/5 of the total streptavidin volume required to the monomer and protease inhibitor solution, mix well and leave on ice in the dark for 20 min (*see* **Note 4**).

7. Repeat the previous step until all of the required streptavidin-fluorochrome solution has been added.

8. Once all the streptavidin has been added, store the tetramer in the dark in a refrigerator at 4 °C.

3.2 Tetramer Staining Protocol

Carry out all procedures at room temperature unless otherwise specified.

1. Acquire single cell suspension of cells and count. Aliquot the cells at 0.5–1 million cells per FACS tube (i.e., per sample including a positive and negative tetramer control sample— *see* **Note 5**).

2. Add 1 ml of TWB to wash the cells and pellet them in a centrifuge at 400 × *g* for 5 min at 4 °C.

3. While the cells are pelleting, defrost a 5 μl (1 mM) aliquot of Dasatinib and add to 50 ml of TWB (1:10,000 dilution), giving a concentration of 100 nM.

4. Pour off the supernatant and remove any additional supernatant using a pipette ensuring that there is no disruption of the cell pellet.

5. Once aspirated, resuspend the cell pellet in 50 μl of TWB, then add 50 μl of the previously prepared 100 nM Dasatinib solution (**step 3**), giving a final concentration of 50 nM of Dasatinib.

6. After the Dasatinib has been added, incubate the cells at 37 °C in 5 % CO_2 for 30 min.

7. While the cells are incubating, change the temperature on the centrifuge to room temperature (~22 °C), which prevents any sudden temperature shock to the cells and improves viability. This incubation time also enables the preparation of the tetramer to be added (see **Note 6**). Add 1 μg of tetramer to maintain a total staining volume of 50 μl. Keep the tetramer on ice and in the dark when not in use, as this prevents tetramer degradation.

8. Post-incubation, pellet the cells (400 × g, 5 min, 22 °C) and discard the supernatant by pouring it off. Then resuspend each tube in 25 μl wash buffer containing 2 μl Trustain and incubate at room temperature for 5 min.

9. Post-incubation add 100 μl of TWB to wash the samples and pellet the cells (400 × g, 5 min, 22 °C).

10. Post-pelleting, discard the supernatant by tipping the solution off and add 50 μl of tetramer to each of the samples. Incubate the samples at 37 °C in 5 % CO_2 for 15 min (see **Note 7**).

11. During this incubation, make up the antibody mastermix to stain the cells in 100 μl in TWB (see **Note 8**) and keep the antibody solution on ice, in the dark, until ready to be used. During this time, the compensation controls should be set up for the flurochromes/antibodies to be used, including a control for the fluorochrome used in the tetramer and an unstained control.

12. Post-incubation, wash the cells in 1 ml TWB and pellet the cells at 400 × g for 5 min at 22 °C. Then pour off the supernatant and resuspend the pellet in 100 μl of the antibody mastermix. Incubate the cells in the refrigerator for 30 min at 4 °C.

13. Lower the temperature of the centrifuge to 4 °C to pellet the cells after the incubation at 400 × g for 5 min at 4 °C

14. Pour off the supernatant and resuspend in 100 μl of TWB. Place the samples at 4 °C in the dark until analysis on a flow cytometer, gating on live single CD8$^+$CD19$^-$CD11b$^-$CD4$^-$Tetramer$^+$ T cells (see **Note 9**). Figure 1 shows the staining using H-2K^d-insulin B_{15-23} Tetramer and T cell receptor transgenic T cells specific for Insulin B_{15-23} peptide.

4   Notes

1. A high quality streptavidin is required for tetramer staining. Ideally a higher concentration of streptavidin should be used to ensure that all the monomers are bound, as free floating monomers will bind to the TCR and make the tetramer staining less

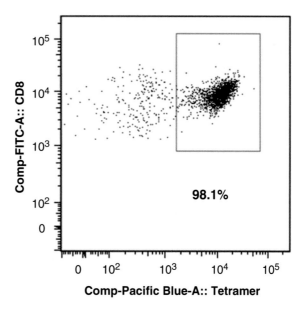

Fig. 1 Flow-cytometric analysis of G9Cα$^{-/-}$ TCR transgenic CD8$^+$ T cells showing staining with FITC-conjugated antibody against CD8 and BV421-conjugated H-2K^d-insulin B$_{15-23}$ peptide tetramer. Percentage shown refers to live CD8$^+$CD4$^-$CD11b$^-$CD19$^-$Tetramer$^+$ T cells

effective. In our work, we identified the Life technologies premium grade APC-streptavidin (S-32362) at 1 mg/ml and also BV421 streptavidin (BioLegend) at 0.5 mg/ml (405225) as good to use for tetramers due to the brightness of the fluorochromes and lower background staining (i.e., less non-specific binding).

2. The antibody mastermix chosen for use will vary depending on the tissue source and additional phenotype information required. In our work, we remove CD19$^+$ B cells, CD4$^+$ T cells, CD11b$^+$ monocytes and macrophages and dead cells. It is important, when devising the phenotyping panel using antibodies to exclude cells that may contribute to nonspecific binding, that possible spectral overlap between the tetramer and any exclusion antibody (i.e., FITC and PE) is considered. Where possible, try to use antibodies on channels that are excited by other lasers to ensure maximum specificity in the staining.

3. If there is a lower concentration of streptavidin-fluorochrome complex, i.e., 0.5 mg/ml, the final volume is multiplied by 2 as twice as much is required to assemble the tetramer.

4. If the volume of streptavidin-fluorochrome solution is too low to add over five incubations, the streptavidin-fluorochrome solution may be diluted in sterile 1× PBS and 1/5 of the total volume (including PBS) to the monomer is added each time.

5. When carrying out tetramer staining it is important to have a positive control, which consists of cells known to bind the tetramer specifically. Cloned T cells or in the case of work with murine models, T cell receptor transgenic cells are particularly useful. A negative control is important, as in any experiment. The positive control will indicate that the tetramer is staining well. The negative control is important for estimating background staining. This background level can then be deducted from the positive staining to give a more accurate result. In this example shown TCR transgenic T cells have been used for the positive control T cells, which recognize the tetramer (H-2K^d presenting Insulin B$_{15-23}$) and the negative control is H-2K^d presenting a minimal binding peptide AYAAAAAAV. This minimal peptide has two amino acids, in this case for H-2K^d (one at position 2 and the other at position 9), that are needed to bind to the MHC.

6. Once the tetramer has been made, it can have a short life so it is best to use the tetramer within 3–4 weeks. Prior to tetramer staining, where possible, a titration should be performed to identify the best concentration of tetramer to use to ensure all the positive control is stained, without increasing nonspecific staining. When titrating tetramer the concentrations we use are 0.25, 0.5, 1, 1.5, and 2 μg.

7. The protocol presented here is a warm tetramer staining method at 37 °C but it is possible to conduct tetramer staining at 4 °C or at room temperature but this will also need to be tested with varying incubation times to ensure optimal tetramer staining.

8. The individual antibodies in the mastermix in this method should be pre-titrated and is used at 100 μl per sample and consists of anti-CD8 FITC, anti-CD4 PeCy7, anti-CD19 PerCpCy5.5, anti-CD11b APC, and Viability dye eFluor 780 made up to 100 μl with TWB per sample.

9. When gating on specific cell populations, any nonspecific binding—dead cells and doublets (formed by cells adhering together) should be gated out, as well as other cell populations, i.e., B cells, monocytes, and macrophages and CD4$^+$ T cells, as shown in the example.

Acknowledgement

This work was supported by a Medical Research Council (grant number G0901155) to FSW and a Diabetes UK PhD studentship to JAP. We acknowledge the NIH Tetramer Core Facility (contract HHSN272201300006C) for provision of MHC | peptide tetramers.

References

1. Altman JD, Moss PA, Goulder PJ et al (1996) Phenotypic analysis of antigen-specific T lymphocytes. Science 274(5284):94–96

2. Wooldridge L, Ekeruche-Makinde J, van den Berg HA et al (2012) A single autoimmune T cell receptor recognizes more than a million different peptides. J Biol Chem 287(2):1168–1177

3. Weichsel R, Dix C, Wooldridge L et al (2008) Profound inhibition of antigen-specific T-cell effector functions by dasatinib. Clin Cancer Res 14(8):2484–2491

4. Wong FS, Karttunen J, Dumont C et al (1999) Identification of an MHC class I-restricted autoantigen in type 1 diabetes by screening an organ-specific cDNA library. Nat Med 5 (9):1026–1031

5. Lieberman SM, Evans AM, Han B et al (2003) Identification of the beta cell antigen targeted by a prevalent population of pathogenic CD8+ T cells in autoimmune diabetes. Proc Natl Acad Sci U S A 100(14):8384–8388

6. Wong FS, Siew LK, Scott G et al (2009) Activation of insulin-reactive CD8 T-cells for development of autoimmune diabetes. Diabetes 58 (5):1156–1164

Methods in Molecular Biology (2016) 1433: 127–134
DOI 10.1007/7651_2015_293
© Springer Science+Business Media New York 2015
Published online: 21 January 2016

Tracking Immunological Responses of Islet Antigen-Specific T Cells in the Nonobese Diabetic (NOD) Mouse Model of Type 1 Diabetes

Terri C. Thayer and F. Susan Wong

Abstract

Tracking autoreactive cells in vivo is important in the study of autoimmune diseases, such as type 1 diabetes. This method provides a model to study the responses of T cells responding to physiologically relevant and organ-specific antigen. Intracellular fluorescent tracers are useful tools to identify adoptively transferred T cells. Firstly, they provide a unique fluorescent signal to distinguish adoptively transferred from endogenous cells. Secondly, cytoplasmic dyes can be used to evaluate proliferation, as the fluorescent intensity is halved with each round of cell division. This provides an important readout to assess cell activation and function.

Keywords: CD8 T cell, CFDA SE, Antigen-specific proliferation, Nonobese diabetic mouse, Type 1 diabetes

1 Introduction

Autoimmune diseases are caused by impaired tolerance to self-antigen. In the case of type 1 diabetes, the immune system specifically recognizes multiple antigens from the pancreatic islets leading to destruction of insulin-producing beta cells. The ability to study immune responses to autoantigens is important for understanding the selection, activation, and regulation of autoimmune effectors in this process.

T cell transgenic mouse models provide an important tool to aid in these endeavors. These systems allow for the generation of mice producing T cells of a particular clone and specificity. There are a number of such mice expressing the T cell receptor (TCR) of autoreactive T cells isolated from the nonobese diabetic (NOD) mouse model of diabetes. Of interest is the $G9C\alpha^{-/-}$ mouse model, as the $CD8^+$ T cells generated recognize a peptide of insulin, an essential autoantigen in autoimmune diabetes, and are potent autoimmune effectors [1]. The ability to monitor these cells in vivo provides the opportunity to study trafficking, cell interactions, activation, and/or regulation during the autoimmune process, as well as presentation of specific autoantigenic peptides by local antigen-presenting cells. For instance, proliferation by islet autoantigen-specific T cells is only seen in unmanipulated NOD mice in the pancreatic lymph nodes, which drain the pancreas, due to endogenous antigen presentation in these local lymph nodes. Although the protocol here uses the insulin-reactive G9 CD8[+] T cells as an example, the procedure could also be applied to other autoantigen-reactive CD8[+] T cells such as NY8.3 [2, 3], which recognizes islet-specific glucose-6-phosphatase catalytic-subunit-related protein (IGRP). In addition, it could also be used for autoreactive CD4[+] T cells such as BDC2.5 [4], a T cell clone that recognizes a peptide of chromogranin A [5].

2   Materials

All reagents should be prepared in a sterile environment appropriate for tissue culture use. Mice are housed in specific-pathogen free facilities.

2.1   Mice

Donor cells are harvested from TCR transgenic mice of interest. G9Cα$^{-/-}$ mice were developed to facilitate the study of the G9C8 CD8[+] T cell, a highly diabetogenic T cell clone upon activation and transfer [6]. The G9C8 T cell receptor is specific for insulin B chain peptide (B15–23) [7]. Young, nonobese diabetic (NOD) mice are used as recipients to study activation of donor cells in an immunologically intact environment.

2.2   Reagents and Supplies

1. Appropriate T cell medium (e.g., RPMI 1640) with 5 % fetal bovine serum (FBS)

2. Necropsy equipment, including scissors and fine forceps

3. Glass Dounce homogenizer, sterilized with alcohol or autoclaved prior to use

4. Deionized water, sterile and stored at room temperature

5. 10× Phosphate-buffered saline (PBS), sterile and stored at room temperature

6. 1× PBS, sterile and stored at room temperature

7. 40 μm cell strainer

8. Magnetic activated Cell Sorting (MACS) Buffer: PBS (phosphate buffered saline), pH 7.2 with 0.5 % bovine serum albumin (BSA) and 2 mM ethylenediaminetetraacetic acid (EDTA) (*see* **Note 1**).

9. Mouse CD8a (Ly-2) microbeads (Miltenyi) for magnetic sorting

10. MACS columns and magnetic separators (*see* **Note 2**)

11. Vybrant® CFDA SE Cell Tracer Kit (Invitrogen); kit components include lyophilized CFDA SE stock (Component A) and high-quality DMSO (Component B) (*see* **Note 3**)

12. Normal saline, sterile and stored at room temperature

13. 1 ml syringes and 27½ gauge needles

14. Mouse restrainer

3 Methods

3.1 Collection of Donor Tissue

1. Donor mice should be culled following institutional guidelines.

2. Spleens from donor mice are collected in 5 ml of T cell medium and stored on ice.

3. Homogenize each spleen in a Dounce homogenizer, rinsing with medium and collecting the single cell suspension in a 50 ml conical tube.

4. Centrifuge cell suspensions at $300 \times g$ for 5 min at 4 °C, and then decant supernatant.

5. Red blood cells are lysed by resuspending cells in 900 µl of sterile deionized water, immediately followed by the addition of 100 µl 10× PBS (*see* **Note 4**).

6. Collect the cell suspension in 5 ml of MACS buffer and pass through a cell strainer to remove debris and obtain a single-cell suspension.

7. Centrifuge cell suspensions at $300 \times g$ for 5 min at 4 °C, and then decant supernatant.

8. Resuspend the pellet in 10 ml of MACS buffer.

9. Count cells to determine total cell number for each sample.

3.2 Magnetic Labeling and Separation

CD8a$^+$ cells will be collected via magnetic separation using anti-mouse CD8a antibody bound to magnetic microbeads. Passing the cells through a column held in a magnetic field will retain labeled cells while unlabeled cells are passed through. Once the column is removed from the magnetic field, positively labeled cells can be collected.

1. Centrifuge cell suspensions at $300 \times g$ for 10 min at 4 °C.

2. Completely remove the supernatant by pipetting or aspiration.

3. Resuspend cells in 90 µl of cold, sterile, and degassed MACS buffer (*see* **Note 1**) per 10^7 total cells, scaling up volumes as needed. If fewer than 10^7 cells, use the volume given for 10^7 total cells.

4. Add 10 µl per 10^7 total cells, again scaling up as needed, of CD8a microbeads provided with the kit. Mix well for even binding and incubate for 15 min at 4 °C.

5. Wash away excess beads with 10 ml of cold MACS buffer. Centrifuge cells at 300 × g for 10 min at 4 °C.

6. Completely remove the supernatant by pipetting or aspiration.

7. Resuspend cells in 500 µl MACS buffer per 10^8 total cells; scale volume as needed.

8. Using the appropriate size column (*see* **Note 2**), set up columns in the magnetic field mounted on a MACS separator in a sterile, tissue culture environment. Place appropriate waste collection tube below column.

9. Rinse column with cold MACS buffer to remove air and prime for separation (*see* **Note 2**).

10. Gently apply cell suspension over column, taking care not to introduce bubbles.

11. When column finishes dripping, wash column with MACS buffer. *See* **Note 2** and manufacturer guidelines for volumes for different column sizes. Rinse the column three times with the appropriate volume.

12. Remove the column from the magnetic field to release bound, labeled cells. Add the appropriate volume of MACS buffer to the column following size guidelines. With the provided plunger, push buffer through column, collecting labeled cells in a sterile conical tube.

13. Centrifuge cell suspensions at 300 × g for 10 min at room temperature.

14. Decant supernatant and resuspend the pellet in 10 ml of 1× PBS.

15. Count cells to determine total cell number for each sample.

3.3 CFDA SE Labeling

1. Pre-warm 1× PBS and Medium containing 5 % FBS in a 37 °C water bath.

2. Thaw an aliquot of 10 mM CFDA SE (*see* **Note 3**). Dilute to 2 µM final concentration in the pre-warmed 1× PBS (*see* **Note 5**).

3. Centrifuge cell suspensions at 300 × g for 10 min at room temperature.

4. Decant supernatant completely and resuspend the pellet at 5×10^6 cells/ml in the pre-warmed PBS containing the 2 µM CFDA SE probe (*see* **Note 6**).

5. Incubate the cells with the probe for 15 min at 37 °C.

6. Centrifuge cell suspensions at $300 \times g$ for 10 min at room temperature.

7. Decant supernatant completely and resuspend the pellet at 5×10^6 cells/ml in the pre-warmed medium with 5 % FBS (*see* **Note 6**).

8. Incubate the cells for 30 min at 37 °C.

9. Centrifuge cell suspensions at $300 \times g$ for 10 min at room temperature.

3.4 Adoptive Transfer of CFDA SE Labeled Cells

1. Wash cells two times in 1× PBS to remove excess dye and serum.

2. Resuspend cells in 10 ml saline and count to determine total cell number for each sample.

3. Centrifuge cell suspensions at $300 \times g$ for 10 min at room temperature.

4. Resuspend cells at 10×10^6 cells per 100 µl in saline to prepare for injection.

5. Draw up cells in a 1 ml syringe and fit with 27½ gauge needle. Remove any air bubbles from the syringe and needle (*see* **Note 7**).

6. In the appropriate specific-pathogen free animal facility, prepare recipient mice for adoptive transfer (*see* **Note 8**).

7. Carefully place mouse in a suitable restrainer.

8. Holding mouse steady, gently rub tail to warm and expose vein.

9. Insert needle into tail vein and inject 100 µl to transfer 10×10^6 labeled cells (*see* **Note 9**).

10. Inject all recipient mice and return to the appropriate facility until harvesting tissue 2–6 days after transfer.

11. Save remaining cells at 4 °C to use as a compensation control when setting up the flow cytometer on the day of analysis.

3.5 Harvesting and Analysis of Transferred Cells

1. Cull recipient mice following institutional guidelines.

2. Collect pancreatic, mesenteric, and para-aortic lymph nodes, or other tissue as needed, from recipient mice. Place tissue in a petri dish with 2 ml T cell medium. Tease open lymph nodes with a 30 gauge needle and fine forceps to release lymphocytes

3. Collect lymphocyte suspensions in a FACS tube.

4. Centrifuge cell suspensions at $300 \times g$ for 10 min at 4 °C.

5. Cells can be labeled with surface and activation markers, including CD8, CD69, CD25, CD62L, or CD44 (*see* **Note 10**).

6. Resuspend cells in 100 µl PBS (with 0.5 % BSA) containing fluorophore-conjugated antibodies.

7. Incubate for 30 min at 4 °C in the dark.

8. Wash cells with 1 ml of PBS (with 0.5 % BSA) and centrifuge at $300 \times g$ for 10 min at 4 °C. Decant supernatant.

9. Resuspend cells in 100 μl PBS (with 0.5 % BSA) or other suitable buffer for flow cytometry.

10. Samples can be analyzed using a flow cytometer capable of detecting the required fluorophores (*see* **Note 11**).

4   Notes

1. If using a powder form of EDTA, the pH of the buffer needs to be checked. The buffer should be sterilized and stored at 4–8 °C. Degassing the buffer under a vacuum or in a sonicator will remove air bubbles, which could potentially block the column and lower cell yield and/or purity.

2. The appropriate column size should be chosen following the guidelines provided by the manufacturer. Choosing the correct column provides sufficient binding space for labeled cells while allowing unlabeled cells to pass freely through the column. This can have a significant impact on yield and purity if the capacity is not sufficient for the cell number. The guidelines will also specify the needed amounts of buffer to wash through the columns.

3. Following kit directions, 10 mM stocks of CFDA SE should be prepared prior to use. One vial of CFDA SE stock (Component A) is to be resuspended in 90 μL of the high-quality DMSO provided (Component B). This can be divided into 5 μl aliquots and stored at −20 °C, protected from the light, until used.

4. Water can be used to lyse red blood cells; however, this requires a quick addition of water immediately followed by the 10× PBS within several seconds. Any delay in the addition of 10× PBS will result in the lysis of white blood cells.

5. The final working concentration of the CFDA SE probe needs to be carefully determined for each cell type and the assay time course. The dye has low toxicity; however the concentration should be tested for any impact on cell viability. The dye can be titrated in order to determine the concentration needed to obtain a bright signal on the flow cytometer used to detect the signal over the experiment timeframe. CFDA SE is a good choice for long-term assays due to its high stability and low toxicity.

6. Cells should be fully resuspended in a single cell suspension to allow for even dye labeling.

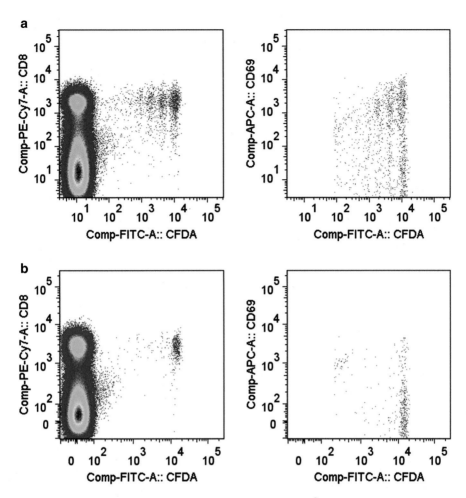

Fig. 1 Analysis of CFDA SE labeled cells. G9 CD8$^+$ T cells (10×10^6) labeled with CFDA SE (2 μM) were transferred intravenously to 6-week-old NOD recipients. After 4 days, pancreatic (**a**) or mesenteric (**b**) lymph nodes were harvested and assessed for the presence of transferred cells. Adoptively transferred cells were identified as CD8$^+$CFDA SE$^+$ cells. CD8$^+$ T cells that had accumulated within the pancreatic lymph nodes (**a**) underwent division, recognizing endogenous antigen presentation. Dividing cells also upregulated CD69, demonstrating activation of cells in the pancreatic lymph nodes (**a**, *right panel*) where the target antigen, insulin, is presented. No proliferation or activation was seen in cells that were collected from the mesenteric lymph node (**b**)

7. All air must be removed from the syringe and sample; air in the blood vessels or heart is fatal.

8. NOD mice, 5–6 weeks of age, are used as recipient mice to track the activity of transferred cells in response to islet-antigen. NOD mice at this age provide the necessary environment for engraftment with sufficient immune activity to release pancreatic autoantigens for the detection of endogenous antigen presentation in the pancreatic lymph nodes.

9. There is no resistance to injection if the needle is properly in the vein.

10. Assessment of markers of activation or regulation can be combined with the proliferation information gained from CFDA SE dilution. CFDA SE excitation is 492 nm and emits at 517 nm. Fluorophores for additional markers need to be chosen carefully as CFDA SE is detected in several channels. A compensation matrix should be validated ahead of time to determine overlap in channels used such as those to detect PE and PerCp.

11. Transferred cells can be distinguished from endogenous cells based on the CFDA SE labeling. As this is a cytoplasmic dye, the fluorescence intensity is halved with each division. The number of division peaks, as well as the percentage of cells in each generation can be used to evaluate the rounds of proliferation achieved (Fig. 1).

Acknowledgement

This work was supported by a Medical Research Council (grant number G0901155) to FSW.

References

1. Wong FS, Siew LK, Scott G, Thomas IJ, Chapman S, Viret C, Wen L (2009) Activation of insulin-reactive CD8 T-cells for development of autoimmune diabetes. Diabetes 58 (5):1156–1164. doi:10.2337/db08-0800

2. Verdaguer J, Schmidt D, Amrani A, Anderson B, Averill N, Santamaria P (1997) Spontaneous autoimmune diabetes in monoclonal T cell nonobese diabetic mice. J Exp Med 186 (10):1663–1676

3. Lieberman SM, Evans AM, Han B, Takaki T, Vinnitskaya Y, Caldwell JA, Serreze DV, Shabanowitz J, Hunt DF, Nathenson SG, Santamaria P, DiLorenzo TP (2003) Identification of the beta cell antigen targeted by a prevalent population of pathogenic CD8+ T cells in autoimmune diabetes. Proc Natl Acad Sci U S A 100 (14):8384–8388. doi:10.1073/pnas. 0932778100

4. Lee LF, Xu B, Michie SA, Beilhack GF, Warganich T, Turley S, McDevitt HO (2005) The role of TNF-alpha in the pathogenesis of type 1 diabetes in the nonobese diabetic mouse: analysis of dendritic cell maturation. Proc Natl Acad Sci U S A 102(44):15995–16000. doi:10. 1073/pnas.0508122102

5. Stadinski BD, Delong T, Reisdorph N, Reisdorph R, Powell RL, Armstrong M, Piganelli JD, Barbour G, Bradley B, Crawford F, Marrack P, Mahata SK, Kappler JW, Haskins K (2010) Chromogranin A is an autoantigen in type 1 diabetes. Nat Immunol 11(3):225–231. doi:10.1038/ni.1844

6. Wong FS, Visintin I, Wen L, Flavell RA, Janeway CA (1996) CD8 T cell clones from young nonobese diabetic (NOD) islets can transfer rapid onset of diabetes in NOD mice in the absence of CD4 cells. J Exp Med 183(1):67–76

7. Wong FS, Karttunen J, Dumont C, Wen L, Visintin I, Pilip IM, Shastri N, Pamer EG, Janeway CA (1999) Identification of an MHC class I-restricted autoantigen in type 1 diabetes by screening an organ-specific cDNA library. Nat Med 5(9):1026–1031. doi:10.1038/12465

Methods in Molecular Biology (2016) 1433: 135–140
DOI 10.1007/7651_2015_294
© Springer Science+Business Media New York 2015
Published online: 21 January 2016

Adoptive Transfer of Autoimmune Diabetes Using Immunodeficient Nonobese Diabetic (NOD) Mice

Evy De Leenheer and F. Susan Wong

Abstract

Studying Type 1 Diabetes (T1D) in the nonobese diabetic (NOD) mouse model can be cumbersome as onset of disease does not usually occur naturally prior to the age of 12–14 weeks and is often restricted to female mice. Furthermore, the onset of disease occurs at random, which makes studying T1D in statistically meaningful cohorts of NOD mice a challenge. Transfer models of T1D into immunodeficient mice, such as NOD SCID mice, allows the study of potential therapeutic interventions in larger cohorts of animals, over shorter periods of time. In this chapter we discuss the adoptive transfer of diabetes into immunodeficient mice on the NOD genetic background that are generally available to the research community.

Keywords: Intravenous cell transfer, CD8 T cells, Adoptive transfer models of type 1 diabetes

1 Introduction

The nonobese diabetic (NOD) mouse is a good model for human type 1 diabetes (T1D), as it develops autoimmune diabetes with some features that are very similar to human type 1 diabetes. CD4 and CD8 T cells, as well as B cells, are important in the development of diabetes. These cells infiltrate the islets of Langerhans and the insulin-producing β cells are damaged and destroyed, with diabetes occurring between 12 and 35 weeks of age [1]. One of the hallmarks of the autoimmune disease is that diabetes can be adoptively transferred by splenocytes from diabetic NOD mice into immunodeficient NOD.SCID mice [2]. Islet antigen-specific CD4 T cells such as BDC2.5 cells recognizing chromogranin A [3–5], CD8 T cells such as G9 cells recognizing insulin B chain 15–23 (B_{15-23}) [6, 7] or NY8.3 cells recognizing islet-specific glucose-6-phosphatase catalytic-subunit-related protein (IGRP) [8–10] can also transfer diabetes. Whereas splenocytes ($1-2 \times 10^7$ cells) extracted and purified from diabetic NOD mice are easily transferred directly ex vivo, in our experience, antigen-specific CD8 T cells taken from T cell receptor (TCR) transgenic mice generated using the TCR from highly diabetogenic CD8 T cell clones require prior activation for reliable adoptive transfer of diabetes. We discuss the protocol of adoptive transfer of autoreactive islet antigen-specific CD8 T cells.

2 Materials

All mice are housed in specific pathogen-free conditions on a 12 h light–dark cycle, were fed autoclaved food and filtered water and kept in cages either individually ventilated or in isolator cages, ventilated with filtered air. All experimental procedures should be performed in accordance with ethically approved institutional protocols for animal research.

2.1 Mice

1. NOD mice are used as donors of bone marrow to generate dendritic cells

2. NOD SCID mice are used for intravenous (IV) cell transfers between the ages of 4 and 6 weeks.

3. CD8 T cell receptor (TCR) G9Cα−/− transgenic mice were generated expressing a specific TCR recognizing insulin B_{15-23}, which was previously cloned from highly diabetogenic G9C8 cells, and are referred to as G9 mice [6].

4. CD8 TCR transgenic mice, recognizing $IGRP_{206-214}$ [10], are available from the Jackson Laboratories and are referred to as 8.3 mice.

5. Donor spleen cells for adoptive transfer are harvested between the ages of 4 and 12 weeks. 8.3 mice are tested for glycosuria and excluded for in vitro stimulation if they are found to be diabetic (blood glucose >13.9 mmol/l).

2.2 Dendritic Cells

1. Bone marrow from 6- to 12-week-old NOD mice are collected and cultured with 1.5 ng/ml GM-CSF (in house supernatant from X63-GM-CSF cells) followed by 1 μg/ml LPS 026:B6 (Sigma) activation to generate mature dendritic cells (DCs).

2.3 Peptides and Reagents

1. G9 CD8 cells are activated with 1 μg/ml insulin B_{15-23} (LYLVCGERG) peptide, supplied as a lyophilized powder from GL Biochem Ltd, Shanghai. The peptide is reconstituted at 5 mg/ml in DMSO (*see* **Note 1**) (10 % of total volume) followed by saline (90 % of total volume). Aliquots are stored at −20 °C.

2. 8.3 CD8 cells are activated with 10 ng/ml $IGRP_{206-214}$ (VYLKTNVFL) peptide, supplied as a lyophilized powder from GL Biochem Ltd, Shanghai. The peptide is reconstituted at 5 mg/ml in DMSO (*see* **Note 1**) (10 % of total volume) followed by saline (90 % of total volume). Aliquots are stored at −20 °C.

3. Complete RPMI medium: RPMI1640 containing 2 mM L-glutamine, 100 U/ml penicillin/streptomycin, 50 μM 2-mercaptoethanol, and 5 % FBS.

4. Magnetic activated cell sorting (MACS) buffer: PBS (phosphate buffered saline) pH 7.2 containing 0.5 % BSA and 2 mM ethylenediaminetetraacetic acid (EDTA).

5. Mouse CD8a-microbeads for magnetic sorting (Miltenyi).

6. Normal saline, sterile and stored at room temperature.

2.4 Equipment

1. Sterile scissors and forceps.

2. Small petri dishes.

3. Glass Dounce homogenizer, sterilized with alcohol or autoclaved prior to use.

4. 25G and 27G needles.

5. 1–10 ml syringes.

6. Mouse restrainer.

7. MACS columns and magnetic separators.

3 Methods

3.1 Generation of Mature Dendritic Cells (DCs)

1. Humanely cull 6–12-week-old female or male NOD mouse according to institutional guidelines.

2. Remove both tibiae and femora using scissors and forceps, ensuring all fur and as much muscle as possible is removed; ensure work takes place in a sterile environment.

3. Spray bones with 70 % ethanol and quickly add to 20 ml complete RPMI medium.

4. Transfer bones to sterile petri dish containing complete RPMI medium, remove both ends of each bone with sterile scissors and flush bone marrow out of bones with a 10 ml syringe fitted with a 25G needle.

5. Withdraw bone marrow suspension using syringe without the needle and expel cells through needle into a fresh 50 ml tube fitted with a 40 μm cell strainer.

6. Spin cell suspension at $400 \times g$ for 5 min at room temperature (RT).

7. Pour off the supernatant and resuspend cells in 20 ml of warm complete RPMI medium

8. Transfer cells into a 75 cm^2 flask (maximum cells from six sets of tibiae and femora per flask).

9. Incubate cells at 37 °C, 5 % CO_2 for 2 h (to remove adherent cells).

10. After incubation remove the supernatant into a new 50 ml tube.

11. Gently wash the flask with 20 ml of complete RPMI medium and add to cell suspension.

12. Using a hemocytometer, count all large cells.

13. Spin cell suspension at $400 \times g$ for 5 min at RT.

14. Resuspend cells at 1×10^6 cells/ml in complete RPMI medium containing 1.5 ng/ml GM-CSF.

15. Add 5 ml cell suspension per well of a 6-well plate.

16. Incubate cells at 37 °C, 5 % CO_2.

17. Change medium every 3 days by removing 2.5 ml of medium and replacing with 2.5 ml complete RPMI medium with 1.5 ng/ml GM-CSF. Be careful not to disturb the cells at the bottom of the well.

18. On day 6–9, add 5 μl of 1 μg/ml lipopolysaccharide (LPS) per well and incubate for 16–18 h for maturation.

19. Harvest cells using cell scraper.

20. Wash cells 3× with complete RPMI medium to remove all LPS and resuspend in 20 ml complete RPMI medium.

21. Irradiate cells (3000 rad) and keep at 37 °C, 5 % CO_2 until ready to use.

3.2 Activation of CD8 Cells

Day 0:

1. Collect spleens from either G9 or 8.3 cells in complete RPMI medium.

2. Homogenize spleens using glass homogenizer.

3. Remove macroscopic tissue debris.

4. Spin cells at $400 \times g$, 5 min, 4 °C.

5. Lyse RBC by adding 900 μl dH_2O immediately followed by 100 μl 10× PBS and 4 ml MACS Buffer.

6. Spin cells at $400 \times g$, 5 min, 4 °C.

7. Resuspend in MACS buffer (10 ml or more depending on number of spleens being processed).

8. Count cells.

9. Spin cells at $400 \times g$, 5 min, 4 °C.

10. Use CD8-microbeads (Miltenyi) to sort CD8$^+$ spleen cells following the manufacturer's guidelines.

11. After separation, spin cells and resuspend cells in 10 ml complete RPMI medium.

12. Count cells.

13. Calculate volume for final concentration of CD8$^+$ cells at 10^6/ml

14. Calculate number of DCs needed (1:20 DC:CD8 ratio or final concentration of DCs is 0.5×10^5/ml)

15. Add CD8$^+$ cells and DCs together, top up to total volume to give 10^6/ml CD8 cells

16. Add insulin B$_{15-23}$ peptide at 1 μg/ml or IGRP$_{206-214}$ peptide at 10 ng/ml, depending on CD8 cell type to be used

17. Add 5 ml cell suspension per well of 6-well plate

18. Incubate for ~24–36 h (after this the cells start to die)

Day 1–2:

1. Cells should have an activated appearance (clusters of blasting cells surrounded by larger single cells)

2. Harvest cells with 1 ml micropipette

3. Rinse each well with 1 ml sterile PBS

4. Spin at $400 \times g$, 5 min, RT

5. Resuspend cells in 10 ml saline

6. Spin at $400 \times g$, 5 min, RT

7. Repeat this 2×; count cells prior to final spin

8. Resuspend cells at $5-10 \times 10^6$/μl PBS, dependent on the number of cells required. For G9 cells, 7×10^6 cells will be sufficient to induce diabetes in 7–10 days. For 8.3 cells, fewer cells may be sufficient.

9. Inject 200 μl cell suspension IV per NOD SCID mouse

Day 5/6:

1. Check mice on a daily basis from day 5/6 onwards for glucose in urine

2. If high for 2 consecutive days, test blood glucose and cull mouse if blood glucose is >13.9 mmol/l (>250 mg/dl).

4 Notes

1. Reconstitution of peptides: do not use PBS instead of saline as this will result in the peptide coming out of solution

Acknowledgements

This work was supported by a Medical Research Council (grant number G0901155) to FSW.

References

1. van Belle TL, Coppieters KT, von Herrath MG (2011) Type 1 diabetes: etiology, immunology, and therapeutic strategies. Physiol Rev 91:79–118

2. Christianson SW, Shultz LD, Leiter EH (1993) Adoptive transfer of diabetes into immunodeficient NOD-scid/scid mice. Relative contributions of CD4+ and CD8+ T-cells from diabetic versus prediabetic NOD.NON-Thy-1a donors. Diabetes 42:44–55

3. Haskins K, McDuffie M (1990) Acceleration of diabetes in young NOD mice with a CD4+ islet-specific T cell clone. Science 249:1433–1436

4. Katz JD, Wang B, Haskins K et al (1993) Following a diabetogenic T cell from genesis through pathogenesis. Cell 74:1089–1100

5. Stadinski BD, Delong T, Reisdorph N et al (2010) Chromogranin A is an autoantigen in type 1 diabetes. Nat Immunol 11:225–231

6. Wong FS, Siew LK, Scott G et al (2009) Activation of insulin-reactive CD8 T-cells for development of autoimmune diabetes. Diabetes 58:1156–1164

7. Wong FS, Visintin I, Wen L et al (1996) CD8 T cell clones from young nonobese diabetic (NOD) islets can transfer rapid onset of diabetes in NOD mice in the absence of CD4 cells. J Exp Med 183:67–76

8. Lieberman SM, Evans AM, Han B et al (2003) Identification of the beta cell antigen targeted by a prevalent population of pathogenic CD8+ T cells in autoimmune diabetes. Proc Natl Acad Sci U S A 100:8384–8388

9. Nagata M, Santamaria P, Kawamura T et al (1994) Evidence for the role of CD8+ cytotoxic T cells in the destruction of pancreatic beta-cells in nonobese diabetic mice. J Immunol 152:2042–2050

10. Verdaguer J, Schmidt D, Amrani A et al (1997) Spontaneous autoimmune diabetes in monoclonal T cell nonobese diabetic mice. J Exp Med 186:1663–1676

Part IV

Evolving Methodologies

Methods in Molecular Biology (2016) 1433: 143–151
DOI 10.1007/7651_2015_286
© Springer Science+Business Media New York 2015
Published online: 21 January 2016

Methylation Analysis in Distinct Immune Cell Subsets in Type 1 Diabetes

Mary N. Dang, Claire M. Bradford, Paolo Pozzilli, and R. David Leslie

Abstract

Epigenetics provides a mechanism in which the environment can interact with the genotype to produce a variety of phenotypes. These epigenetic modifications have been associated with altered gene expression and silencing of repetitive elements, and these modifications can be inherited mitotically. DNA methylation is the best characterized epigenetic mark and earlier studies have examined DNA methylation profiles in peripheral blood mononuclear cells in disease. However, any disease-related signatures identified would just display differences in the relative abundance of individual cell types as each cell subset generates a unique methylation profile. Therefore is it important to identify cell- or tissue-specific changes in DNA methylation, particularly in autoimmune diseases such as type 1 diabetes.

Keywords: Type 1 diabetes, DNA methylation, Cell isolation, Flow cytometry, Magnetic-activated cell sorting, DNA extraction, Illumina HumanMethylation450K BeadChip

1 Introduction

DNA methylation is the most characterized epigenetic modification and is essential for regulating the expression of mammalian genes. It involves the addition of a methyl group to the cytosine in a CpG dinucleotide [1, 2]. DNA methylation has been shown to be tissue-specific [3] and stable for at least 3 years [4]. Epigenetics has been studied in autoimmune diseases such as systemic lupus erythematosus and rheumatoid arthritis as genetic factors cannot fully account for development of these diseases [5]. For example, in type 1 diabetes (T1D), concordance rates in monozygotic twins is ~50 % [6] suggesting a role for non-genetic factors.

We have previously performed an epigenome-wide association study using CD14+ monocytes from T1D-discordant monozygotic twin pair with the Illumina HumanMethylation27K BeadChip [7]. Following on from this, we have extended the study by also isolating CD19+ B cells and CD4+ T cells and the classical subset of monocytes (CD14+CD16−) from monozygotic twin pairs discordant for T1D. Buccal samples were also collected as they have been shown to have its own unique methylome [8]. The PBMC layer was harvested through the use of Percoll, a gradient density medium, and the different cell populations were sorted using magnetic-activated cell sorting (MACS). MACS separates different cell populations through use of magnetic beads coated with antibodies against the surface antigen of interest. The cells are then passed through a column in a magnetic field and the cells of interest are retained in the column for a pure population. MACS has been shown to obtain high purities of isolated cell populations and subsequent use of magnetic beads on a single sample did not significantly affect gene expression [9]. The purity of each cell type collected was then assessed using flow cytometry and subsequently, DNA was extracted from the cells and buccal samples. The samples were treated with sodium bisulfite and hybridized onto the beadchips for DNA methylation profiling using Illumina HumanMethylation450K BeadChip. The Illumina Infinium HumanMethylation BeadChip is an example of an array based technology for DNA methylation profiling [10]. Here, we have demonstrated that different cell types (CD19$^+$, CD14$^+$CD16$^-$, and CD4$^+$) can be isolated from a single blood sample with cell purity over 95 % for epigenetic profiling.

2 Materials

2.1 Cell Isolation

1. 1 M Trisodium Citrate (TNC): Dissolve 29.4 g trisodium citrate in 100 mL of water through a filter.

2. Percoll 1.078 g/mL: 300 mL Percoll (GE Healthcare), 24 mL 10× PBS, 216 mL 1× PBS, 13.8 mL human serum albumin, 7.2 mL 1 M TNC. Store at 4 °C.

3. PBS, 2 mM EDTA (Buffer 2): Add 2 mL 0.5 M EDTA to 500 mL 1× PBS.

4. RPMI: Add 5 mL penicillin streptomycin and 5 mL human serum albumin to 500 mL RPMI 1640 Medium, Gluta-MAX™, HEPES (Invitrogen).

5. BD Vacutainer® Sodium Heparin tube with Green Conventional Closure, 10 mL (BD).

6. LS, LD and MS columns (Miltenyi Biotech).

7. MACS stand and separator (Miltenyi Biotec).

8. Pre-separation filters (Miltenyi Biotech).

9. CD19, CD4, CD16, and CD14 MicroBeads, human (Miltenyi Biotech).

2.2 Flow Cytometry

1. FITC conjugated monoclonal mouse anti-human CD14, clone MφP9 (BD Biosciences).

2. PE conjugated monoclonal mouse anti-human CD16, clone B73.1/leu11c (BD Biosciences).

3. PerCP-Cy5.5 conjugated monoclonal mouse anti-human CD64, clone 10.1 (BD Biosciences).

4. PE-CY7 conjugated monoclonal mouse anti-human CD45, clone HI30 (Invitrogen).

5. FITC conjugated monoclonal mouse anti-human CD4, clone M-T466 (Miltenyi Biotec).

6. PE conjugated monoclonal mouse anti-human CD19, clone LT19 (Miltenyi Biotec).

7. Falcon® 5 mL Round Bottom Polystyrene Test Tube (Corning).

8. Anti-Mouse Ig, κ/Negative Control (FBS) Compensation Particles Set (BD Bioscience).

9. FITC conjugated monoclonal mouse anti-human Ig, κ light chain, clone TB28-2 (BD Biosciences).

10. PE conjugated monoclonal mouse anti-human Ig, κ light chain, clone TB28-2 (BD Biosciences).

11. PerCP-Cy™5.5 conjugated mouse IgG1 κ isotype control (BD Biosciences).

12. PE-Cy® 7 conjugated mouse IgG1 (Invitrogen).

13. FITC conjugated mouse IgG1 (Miltenyi Biotech).

14. PE conjugated mouse IgG1 (Miltenyi Biotech).

15. BD FACSCanto II Flow Cytometer (BD).

2.3 DNA Extraction and Sodium Bisulfite Treatment

1. QIAamp DNA Blood Mini Kit (250) (Qiagen).

2. RNAProtect (Qiagen).

3. Qubit fluorometer (Invitrogen).

4. Qubit dsDNA assay kit (Invitrogen).

5. Qubit assay tubes (Invitrogen).

6. Gentra Puregene Buccal Cell Kit (100) (Qiagen).

7. EZ DNA Methylation kit (Zymo Research).

2.4 Agarose Gel Electrophoresis

1. Gel tank (Bio-Rad).

2. UltraPure agarose (Invitrogen).

3. TBE buffer 10×: Dissolve 108 g tris base, 51 g boric acid, and 40 mL 0.5 M EDTA into 500 mL distilled water.

4. Ethidium bromide solution (Sigma Aldrich).

3 Methods

Keep Buffer 2 and cells on ice unless otherwise specified. Centrifuge sample with brakes unless otherwise specified.

3.1 Peripheral Blood Mononuclear Cell (PBMC) Extraction

1. Collect 50 mL of blood in sodium heparin tubes and dilute 1:1 with RPMI. Leave the sample rolling overnight at room temperature (*see* **Note 1**).

2. Carefully pipette 20–25 mL of the diluted blood onto 12.5 mL Percoll 1.078 g/mL in four 50 mL tubes.

3. Centrifuge the sample for 20 min, at $800 \times g$, with no brake at 20 °C. After centrifugation, the sample will split into different layers (Fig. 1).

4. Remove most of the plasma layer and harvest the PBMC ring into two 50 mL tubes. Wash with the PBMC ring with Buffer 2, centrifuge at $550 \times g$ for 8 min at 20 °C.

5. Pool the cell pellets into one 50 mL tube and fill the tube with Buffer 2. Centrifuge at $550 \times g$ for 8 min at 4 °C.

6. Resuspend the cells in 10 mL Buffer 2. With an aliquot of cell solution, mix 1:1 with trypan blue. Count cells using a haemocytometer.

7. Add up to 45 mL of Buffer 2 to the 50 mL tubes and spin at $550 \times g$ for 6 min at 4 °C.

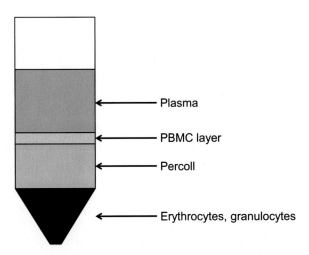

Fig. 1 Extraction of PBMCs on a Percoll gradient. PBMCs are separated from blood using centrifugation on Percoll. The diluted blood is overlayed onto Percoll in a 50 mL Falcon tube. Centrifugation will result in the formation of different layers. The different layers that will be observed from *top* to *bottom* are as follows: plasma, PBMC layer, Percoll, and erythrocytes and granulocytes. The PBMC layer is then harvested and washed

3.2 *Isolating B Cells*

1. Resuspend the cell pellet in 80 μL Buffer 2 per 10 million cells. Then add 20 μL of CD19 MB per 10 million cells.

2. Incubate for 15 min at 4 °C. Add 2 mL of Buffer 2 to cells and spin at 300 × *g* for 6 min at 4 °C.

3. Whilst waiting for the samples to finish spinning, wash an LS column with 3 mL Buffer 2 through a pre-separation filter.

4. Resuspend the cells in 500 μL of Buffer 2 and apply the cell suspension to the column through the pre-separation filter.

5. Add 3 mL of Buffer 2 to the column three more times (*see* **Note 2**). Remove the column from the magnet and add 5 mL of Buffer 2 to the column.

6. Use the plunger to flush out the purified cells. Centrifuge the flow through and purified cells at 300 × *g* for 6 min at 4 °C. Wash an MS column with 500 μL of Buffer 2.

7. Decant the supernatant from the CD19⁻ tube and leave on ice to one side.

8. Resuspend the CD19⁺ fraction in 500 μL of Buffer 2 and add to the MS column (*see* **Note 3**).

9. Add 500 μL of Buffer 2 to column three times. Remove column from the magnet and add 2 mL of Buffer 2 to the column.

10. Use the plunger to flush out the purified cells and perform a cell count.

11. Discard the flow through (second CD19⁻ fraction).

3.3 *Isolating Monocytes*

1. Resuspend the remaining cell pellet to one side (first CD19⁻ fraction) in 50 μL Buffer 2 per 50 million cells. Add 50 μL of CD16 MB per 50 million cells.

2. Incubate for 30 min at 4 °C. Add 2 mL of Buffer 2 to cells. Spin at 300 × *g* for 6 min at 4 °C.

3. Wash an LD column with 2 mL Buffer 2. Resuspend the cells in 500 μL of Buffer 2 and apply the cell suspension to the column.

4. Add 1 mL of Buffer 2 to column twice.

5. Collect the flow through and perform a cell count. Then spin at 300 × *g* for 6 min at 4 °C.

6. Resuspend the pellet in 80 μL Buffer 2 per 10 million cells. Add 20 μL of CD14 MB per 10 million cells.

7. Incubate for 15 min at 4 °C. Add 2 mL of Buffer 2 to cells. Spin 300 × *g* for 6 min at 4 °C.

8. Wash an MS column with 500 μL Buffer 2. Resuspend the cells in 500 μL of Buffer 2 and apply the cell suspension to the column.

9. Add 500 μL of Buffer 2 to column three times. Remove the column from the magnet and add 2 mL of Buffer 2 to column.

10. Use the plunger to flush out the purified cells and perform a cell count.

11. Collect flow through and perform a cell count. Spin at $300 \times g$ for 6 min at 4 °C.

3.4 Isolating T Cells

1. Resuspend the cell pellet in 80 μL Buffer 2 per 10 million cells. Add 20 μL of CD4 MB per 10 million cells.

2. Incubate for 15 min at 4 °C. Add 2 mL of Buffer 2 to cells. Spin at $300 \times g$ for 6 min at 4 °C.

3. Wash MS column with 500 μL Buffer 2. Resuspend cells in 500 μL of Buffer 2. Apply cell suspension to the column.

4. Add 500 μL of Buffer 2 to column three times. Remove the column from the magnet and add 2 mL of Buffer 2 to the column.

5. Use the plunger to flush out the purified cells. Perform a cell count.

3.5 Flow Cytometry

1. Add an aliquot from each cell type (CD19$^+$, CD14$^+$CD16$^-$, and CD4$^+$) to separate 5 mL FACS tubes. Spin down at $300 \times g$ for 6 min at 4 °C (*see* **Note 4**).

2. Resuspend in 100 μL of Buffer 2.

3. Add antibodies (Table 1) (*see* **Note 5**). Vortex tubes and incubate for 10 min at 4 °C.

4. Wash with 2 mL Buffer 2 then centrifuge at $300 \times g$ for 6 min at 4 °C.

5. Resuspend in 500 μL Buffer 2.

6. Set up single color controls with the compensation beads set by adding 100 μL of Buffer 2 into a new tube for each staining

Table 1
Antibody panel for the individual cell types

Monocytes		
CD14	FITC	20 μL
CD16	PE	20 μL
CD64	PerCP-Cy5.5	5 μL
CD45	PE-CY7	5 μL
T cells		
CD4	FITC	10 μL
B cells		
CD19	PE	10 μL

Table 2
Setting up the FMO controls for analyzing monocytes

	FITC	PE	PerCP-Cy5.5	PE-CY7
FITC	–	✓	✓	✓
PE	✓	–	✓	✓
PerCP-Cy5.5	✓	✓	–	✓
PE-CY7	✓	✓	✓	–

The tubes are represented by the first columns (*bold*). The rest of the columns specify which antibody to add in a tube. For example, the FITC FMO will include the PE, PerCP-Cy5.5, and PE-CY7 but not the FITC antibody

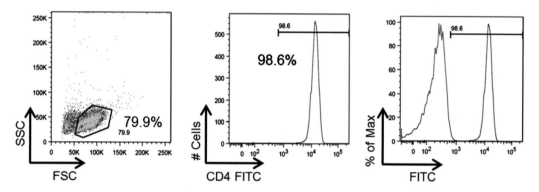

Fig. 2 Flowjo analysis of CD4⁺ cells. Representative FACS results for purified CD4⁺ cells. CD4⁺ cells are stained with FITC conjugated antibodies (*dark blue*). 10,000 events are recorded in the initial gate. Mouse anti-IgG FITC antibody is also used to stain the cells to establish a negative control (*light blue*). Purity for all samples will be over 95 %

antibody. Then add a drop of the negative control and anti-mouse Ig K beads to each tube. Add the same volume of antibodies indicated in Table 1 to the tubes.

7. For setting up isotype controls, use the same volume of antibodies as in Table 1.

8. For setting up fluorescence-minus-one (FMOs) controls, *see* Table 2.

9. Use the BD Canto II instrument to assess the purity of each cell type (Fig. 2).

3.6 DNA Extraction and Sodium Bisulfite Treatment

1. Split each isolated cell type into 2 eppendorf tubes.

2. Spin all at 14,000 rpm for 2 min at RT.

3. Add 200 μL PBS and 200 μL AL lysis buffer to one pellet for DNA extraction, vortex then store at −20 °C.

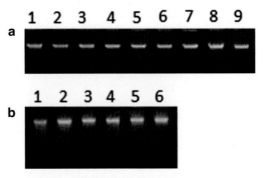

Fig. 3 Quality control for DNA extraction. DNA is measured using a Qubit instrument and 100 ng DNA from purified cells and buccal samples are run on a 2 % agarose gel to assess any degradation. The gels are viewed using a UV gel imager. (**a**) DNA from purified cells, *lanes 1, 4,* and *7* are DNA from CD14$^+$CD16$^-$ cells. *Lanes 2, 5,* and *8* are DNA from CD4$^+$ cells. *Lanes 3, 6,* and *9* are DNA from CD19$^+$ samples. (**b**) DNA from buccal brushes from different individuals

4. Resuspend the remaining pellet in 300 μL RNAprotect, vortex and store at −80 °C for future RNA extraction.

5. Extract DNA from cells using QIAamp DNA Blood Mini Kit.

6. Measure DNA concentration using Qubit.

7. Run 100 ng DNA on a 2 % agarose gel and visualize on a UV transilluminator (Fig. 3).

8. Extract DNA from the buccal brushes using the Gentra Puregene Buccal cell kit according to manufacturer's instructions. Measure the DNA concentration with the Qubit instrument.

9. 500 ng DNA is ready to be sent to a genome center to be treated with sodium bisulfite and then hybridized onto the Infinium HumanMethylation450K BeadChip.

4 Notes

1. Samples are taken at any point in the day and therefore are left rolling overnight in order to standardize the protocol.

2. Wait for the 3 mL of Buffer 2 to flow through the column completely before the addition of another 3 mL.

3. There is no need for pre-separation filters from then on. Sorting cells from PBMCs may clog up the column while positively selecting for CD19 cells.

4. Whilst decanting the supernatant from the 5 mL FACS tubes, blot onto absorbent paper.

5. Sample gating and background signal were determined using unstained, isotype, and fluorescence-minus-one (FMO) controls.

Acknowledgement

This work was supported by JDRF and BLUEPRINT EU-FP7.

References

1. Laird PW (2003) The power and the promise of DNA methylation markers. Nat Rev Cancer 3:253–266

2. Bernstein BE, Meissner A, Lander ES (2007) The mammalian epigenome. Cell 128:669–681

3. Rakyan VK, Down TA, Thorne NP et al (2008) An integrated resource for genome-wide identification and analysis of human tissue-specific differentially methylated regions (tDMRs). Genome Res 18:1518–1529

4. Beyan H, Down TA, Ramagopalan SV et al (2012) Guthrie card methylomics identifies temporally stable epialleles that are present at birth in humans. Genome Res 22:2138–2145

5. Dang MN, Buzzetti R, Pozzilli P (2013) Epigenetics in autoimmune diseases with focus on type 1 diabetes. Diabetes Metab Res Rev 29:8–18

6. Huber A, Menconi F, Corathers S, Jacobson EM, Tomer Y (2008) Joint genetic susceptibility to type 1 diabetes and autoimmune thyroiditis: from epidemiology to mechanisms. Endocr Rev 29(6):697–725

7. Rakyan VK, Beyan H, Down TA et al (2011) Identification of type 1 diabetes-associated DNA methylation variable positions that precede disease diagnosis. PLoS Genet 7, e1002300

8. Lowe R, Gemma C, Beyan H et al (2013) Buccals are likely to be a more informative surrogate tissue than blood for epigenome-wide association studies. Epigenetics 8:1–10

9. Lyons PA, Koukoulaki M, Hatton A, Doggett K, Woffendin HB, Chaudhry AN, Smith KG (2007) Microarray analysis of human leucocyte subsets: the advantages of positive selection and rapid purification. BMC Genomics 8:64

10. Rakyan VK, Down TA, Balding DJ et al (2011) Epigenome-wide association studies for common human diseases. Nat Rev Genet 12:529–541

Methods in Molecular Biology (2016) 1433: 153–158
DOI 10.1007/7651_2015_288
© Springer Science+Business Media New York 2015
Published online: 13 December 2015

Fluorescence In Situ Hybridization with Concomitant Immunofluorescence in Human Pancreas

Jody Ye and Kathleen M. Gillespie

Abstract

The ability to identify the presence of non-host cells in human pancreas with concomitant characterization of cell phenotype is particularly important to facilitate studies of transplantation and microchimerism resulted from pregnancy. The steps involved in processing tissue for fluorescence in situ hybridization (FISH) can however remove epitopes that are crucial for immunofluorescence and antigen retrieval strategies for immunofluorescence can negatively influence FISH. We describe a robust method to analyze X/Y chromosome constitution and cell phenotype simultaneously on the same pancreatic tissue section.

Keywords: Fluorescence in situ hybridization, Transplantation, Microchimerism

1 Introduction

In this chapter, the technique used to detect the X and Y chromosomes in sex-mismatched cells in pancreas (perhaps as a result of a pancreas or bone marrow transplant or maternal fetal cell transfer) and simultaneously determine the cell phenotype on the same cell is described. FISH employs fluorescently conjugated DNA probe to hybridize the centromeric regions of the X/Y chromosomes. Commercially available kits such as Vysis CEP X/Y Direct labeled fluorescent DNA probe kit are often used. Briefly, the alpha satellite sequence of the X chromosome and satellite III at the Yq12 of the Y chromosome contain highly repetitive tandem repeats. Binding to these repeats amplifies the FISH signal. In order to label immunomarkers concomitantly with FISH, DNA targets and protein epitopes must be retrieved under the same condition, for example the same heat-induced antigen retrieval (HIAR) buffer and condition, as illustrated in Fig. 1. This article describes a simple way of combining FISH and immunofluorescence on paraffin-embedded tissues.

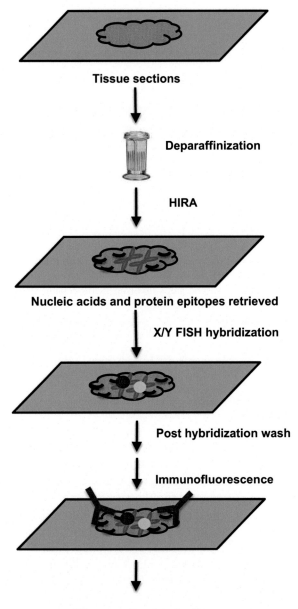

Fig. 1 Schematic illustration of concomitant X/Y chromosome fluorescence in situ hybridization and immunofluorescence on paraffin-embedded tissue sections. *HIAR* heat induced antigen retrieval; *black lines* indicate protein epitopes; *red dot* represents centromere of X chromosome and *green dot* represents centromere of Y chromosome; *dark blue* highlights chromosomes; *light blue* represent antibodies

2 Materials

2.1 Tissue Preparations

Cut formalin fixed paraffin-embedded human pancreas blocks at 4 µM per section and mount on positively charged slides.

2.2 Reagent and Materials for Fluorescence In Situ Hybridization

1. Xylene.

2. Vysis CEP X SpectrumOrange/Y SpectrumGreen Direct labeled fluorescent DNA probe kit (30-161050 and 32-161050, Abbott Molecular).

3. 1× sodium citrate buffer (10 mM, pH 6.3): dissolved 0.294 g tri-sodium citrate in 900 ml water, adjust pH to 6.0, bring total volume to 1 L and then add 0.5 ml Tween 20.

4. 20× saline sodium citrate (SSC) buffer (pH 5.3): dissolve 175.3 g sodium chloride and 88.2 g tri-sodium citrate in 800 ml deionized water (dH$_2$O), adjust pH to 5.3, and bring up the volume to 1 L. Solution can be filtered through a 0.45 µM filtration unit or autoclaved before storing at room temperature for up to 6 months.

5. Ethanol wash solutions: prepare v/v dilutions of 70 %, 85 %, and 100 % using 100 % ethanol in dH$_2$O.

6. 0.1 % Tween 20, 2× SSC wash buffer: dilute 20× SSC buffer (pH 5.3) in dH$_2$O at 1:10, adjust pH to 7.0–7.5, and add 0.1 % v/v pure Tween 20 solution to the buffer.

7. 0.4× SSC wash buffer: dilute 20× SSC buffer (pH 5.3) in dH$_2$O at 1:50 and adjust pH to 7.0–7.5.

8. Denaturing solution: 70 % formamide in 2× SSC solution (without Tween 20) (for separate denaturation and hybridization).

9. Slide warmer (for separate denaturation and hybridization).

10. Coverslips.

11. Rubber cement.

12. Digital slide hybridizer (for co-hybridization only).

13. Water bath (2 if performing separate denaturation and hybridization).

14. Microwave.

15. Coplin jars.

16. Digital thermometer.

17. Slide chamber.

18. Heat stable container for antigen retrieval.

2.3 Reagent for Immuno-fluorescence

1. Normal blocking serum.

2. 0.1 % Triton-X PBS antibody dilution reagent.

3. 1× PBS wash buffer.

4. Primary and fluorescent secondary antibodies.

5. VECTASHIELD DAPI mounting media.

3   Methods

The goal is to achieve optimum DNA target and protein epitope exposure under the same pretreatment condition.

3.1   Deparaffinization and Heat-Induced Antigen Retrieval

This procedure allows a maximum of five slides to be processed simultaneously (**Note 1**).

1. Immerse slides in xylene for 5 min.

2. Repeat in a second xylene wash.

3. Transfer slides to a 100 %, 85 %, and 70 % ethanol series; incubate for 1 min each.

4. Wash in dH_2O water.

5. Carefully preheat 10 mM sodium citrate buffer (pH 6.0) in microwave, bringing to the boil; quickly immerse slides in sodium citrate buffer and continue to heat up at full power ~850 W for at least 10 min (**Note 4**).

6. After antigen retrieval, slides are cooled down slides to room temperature in sodium citrate buffer.

7. Subsequently, slides are dehydrated in 70 %, 85 %, and 100 % ethanol series for 1 min each and air-dried on bench.

8. Slides can be stored overnight at room temperature but best to proceed with in situ hybridization straight after.

3.2   Fluorescence In Situ Hybridization (Co-hybridization)

1. Preheat digital slide hybridizer to 73 °C and water bath to 42 °C, this step requires accurate temperature settings and temperature cannot fluctuate more than 1 °C.

2. Depending on tissue size, pipette appropriate amount of FISH ready-to-use probe mixture onto a coverslip. As a general guidance, 10 μl FISH probe is used on a 2 cm^2 tissue section.

3. Reverse coverslip and place it on top of the tissue, make sure that the tissue is totally immersed in FISH probe, squeeze out any air bubbles.

4. Seal coverslip with rubber cement. From now on slides need to be protected from light, and incubated in slide hybridizer at exactly 73 °C for 10 min. This step allows double stranded DNA duplex to denature in the presence of single stranded FISH probe.

5. Immediately slides are transferred to a humidified slide chamber, allowing probes to re-hybridize with DNA in water bath at 42 °C overnight.

3.3 Fluorescence In Situ Hybridization (Separate Denaturation and Hybridization)

1. Pre-warm the water bath 1 to 73 °C, water bath 2 to 42 °C, and pre-warm the slide warmer to 45–50 °C.

2. Fill denaturation solution (70 % formamide in 2× SSC pH 7.0–8.0) into a Coplin jar and place it in the 73 °C water bath to warm up.

3. Denature specimen by immersing the prepared slides in the denaturation solution at 73 ± 1 °C for 5 min. Do not denature more than five slides.

4. Using forceps, remove slides from the denaturation solution and immediately place into 70 % ethanol, agitate gently, and allow slides to stand for 1 min.

5. Transfer slides to 85 % ethanol for 1 min and 100 % ethanol prior to FISH probe hybridization.

6. Place slides on a 45–50 °C slide warmer no more than 2 min.

7. Apply FISH probe solution on tissue, then place coverslip on top, seal with rubber cement.

8. Transfer slides to slide chamber, cover from light, and incubate in water bath 2 at 42 °C overnight.

3.4 Post Hybridization Wash

1. Preheat water bath to 73 ± 0.5 °C.

2. Preheat 0.4× SSC solution to exactly 73 ± 0.5 °C in a Coplin jar in water bath.

3. Prepare 2× SSC, 0.1 % Tween 20 solution in a Coplin jar.

4. Gently remove coverslip from tissue and immerse slides in 0.4× SSC solution for exactly 2 min. Gently agitate Coplin jar between each minute.

5. Transfer slides to 2× SSC, 0.1 % Tween 20 solution and incubate for 1 min, gently agitate for 30 s.

3.5 Immuno-fluorescence

1. At this point, immunofluorescence can be performed.

2. Block nonspecific binding with 3 % serum from appropriate species (the same species as secondary antibody was raised in) diluted in PBS buffer for 1 h at room temperature.

3. Incubate slides with desired primary antibodies at room temperature for 2 h or overnight at 4 °C.

4. Wash slides in PBS wash buffer twice for 3 min each with gentle agitation.

5. Incubate slides with desired fluorescent secondary antibodies at room temperature for 1 h.

6. PBS wash as described in step 4.

7. Counterstain and mount tissues in VECTASHIELD DAPI mounting media.

8. Seal coverslip with nail varnish.

**3.6 Microscopic
Examination**

1. FISH and immunofluorescence signals can be visualized under regular and confocal microscope preferentially at 60× magnification lens or above.

2. Slides can be stored at −20 °C up to a year.

4 Notes

1. FISH can be performed either as co-hybridization or as separate denaturation and hybridization.

2. Deparaffinizing too many slides simultaneously can reduce the incubation temperature and inferior results.

3. At this stage slides can be stored in deionized water for less than 1 h.

4. Ten minutes are required to break up formalin-induced cross-link between peptides. Depending on target epitopes, the length of HIAR can vary between 10 and 20 min. For abundant antigens in islets such as insulin and glucagon, a 10-min antigen retrieval is sufficient to obtain strong signal. For proliferative markers such as Ki67, a 20-min antigen retrieval is recommended to achieve optimal staining.

Methods in Molecular Biology (2016) 1433: 159–167
DOI 10.1007/7651_2015_290
© Springer Science+Business Media New York 2015
Published online: 13 December 2015

Laser Capture and Single Cell Genotyping from Frozen Tissue Sections

Thomas Kroneis, Jody Ye, and Kathleen Gillespie

Abstract

There is an increasing requirement for genetic analysis of individual cells from tissue sections. This is particularly the case for analysis of tumor cells but is also a requirement for analysis of cells in pancreas from individuals with type 1 diabetes where there is evidence of viral infection or in the analysis of chimerism in pancreas; either post-transplant or as a result of feto-maternal cell transfer.

This protocol describes a strategy to isolate cells using laser microdissection and to run a 17plex PCR to discriminate between cells of haplo-identical origin (i.e., fetal and maternal cells) in pancreas tissue but other robust DNA tests could be used. In short, snap-frozen tissues are cryo-sectioned and mounted onto membrane-coated slides. Target cells are harvested from the tissue sections by laser microdissection and pressure catapulting (LMPC) prior to DNA profiling. This is based on amplification of highly repetitive yet stably inherited loci (short tandem repeats, STR) as well as the amelogenin locus for sex determination and separation of PCR products by capillary electrophoresis.

Keywords: Laser microdissection and pressure catapulting, Pancreatic cells, Diabetes, Tissue

1 Introduction

There has been renewed interest in the study of the pancreas in type 1 diabetes including detailed descriptions of the cells constituting insulitis [1, 2] as well as studies aimed at detecting evidence of viral infection [3, 4], inflammatory mediators [5], and beta cell stress [6]. Such studies have increasingly been enabled by access to high quality autopsy pancreas tissues through the Network of Donors (nPOD; http://www.jdrfnpod.org/). One requirement is the recovery of high quality DNA from individual cells for further DNA investigation, for instance confirmation that a cell, or cells, in a transplanted pancreas originates from the recipient (chimerism) [7] or mother (maternal microchimerism) [8, 9]. We have developed a technique to isolate individual cells from frozen human pancreas for downstream DNA analysis. The DNA test used here generated a profile of 16 polymorphic loci and one sex determining locus [10, 11] but any robust DNA test could be used for DNA analysis, for instance viral DNA tests.

A note of caution; our experiments to date showed that the protocol described here does not work optimally on

1. Formalin fixed paraffin embedded tissue: this is possibly because formalin irreversibly cross-links proteins via the amino groups to preserve the structural integrity of the cells. Isolation of nucleic acids is however impaired by both the paraffin wax and the cross-linking process.

2. Tissue that has been pretested using fluorescence in situ hybridization (FISH) to stain specific DNA sequences.

Further optimisation is required in both these situations.

2 Materials

2.1 Materials for Tissue Preparation

1. Cryostat.
2. Poly ethylene terephthalate (PET) membrane slide.
3. UV light source (254 nm) (i.e., in a tissue culture hood or a DNA cross-linker).
4. 100 % acetone.

2.2 Materials for DAPI Staining

1. 10× phosphate-buffered saline (PBS, pH 7.3): dissolve 2.586 g potassium dihydrogen phosphate (KH_2PO_4), 29.01 g disodium hydrogen phosphate (Na_2HPO_4), and 90.06 g sodium chloride (NaCl) in 900 mL Milli-Q-water. Use a magnet stirrer to dissolve the salts. When all salts are dissolved, top up solution to 1 L and store the solution in a glass flask at RT for up to 6 months.
2. 1× PBS: dilute from 10× PBS buffer, i.e., 100 µL 10× PBS plus 900 µL Milli-Q water (*see* **Note 1**).
3. Ethanol wash solutions: prepare v/v dilutions of 70, 85, and 100 % using 100 % ethanol in purified water.
4. VECTASHIELD mounting medium with DAPI, H-1200 (Vector Laboratories, UK).
5. Coverslips (22 × 22 mm).
6. Coplin jars.

2.3 Materials for LMPC

1. 1× PBS.
2. PCR-clean tubes (200 µL and 1.5 mL).
3. Plastic centrifugation tubes (50 and 15 mL).
4. Dedicated pipette set and filter pipette tips (20, 200, 1000 µL).
5. PCR-grade water.
6. Mini Centrifuge.

2.4   Materials for Cell Lysis

1. 5× Lysis Buffer: 10 mM Tris-HCl, 6 mM Ca(OAc)$_2$: Dissolve 1.7618 g of Ca(OAc)$_2$ per 10 mL of nuclease-free water to obtain a 1 M calcium acetate stock solution. Mix 6 μL of 1 M calcium acetate stock solution and 10 μL 1 M Tris-HCl (pH 7.4) with 984 μL of nuclease-free water and pulse vortex for 15 s. Aliquot the 5× Lysis Buffer and store at −20 °C.

2. Proteinase K storage buffer: 10 mM Tris-HCL, 6 mM Ca(OAc)$_2$, glycerol 40 vol.%: mix 6 μL of 1 M calcium acetate stock solution, 10 μL 1 M Tris-HCl, 400 μL glycerol and 584 μL Milli-Q water.

3. 5× Lysis Enzyme: dissolve 20 mg proteinase K in 1 mL of the Protinase K storage buffer, mix well, aliquot and store at −20 °C.

2.5   Materials for DNA Profiling

1. PCR-clean tubes (200 μL and 1.5 mL).

2. Filter pipette tips (10, 200, 1000 μL).

3. Milli-Q water.

4. Centrifuge for a short spin at max 2000 × g.

5. PowerPlex® 17 ESX System Kit (cat. no. DC6721, Promega, *see* **Note 2**): 2800 M Control DNA (cat. no. DD710A), CC5 Internal Lane Standard500 (cat. no. DG152A), PowerPlex® 17 10× Primer Pair Mix (cat. no. DK149B), PowerPlex® ESX 17 Allelic Ladder Mix (DG150B), PowerPlex® ESX 5× Master Mix (DP102A), Water, Amplification Grade (DW099A).

6. Matrix Standards (*see* **Note 3**).

7. Thermo cycler.

8. Capillary analyzer (e.g., 3730 DNA Analyzer).

9. 96-well plates.

10. Sealing foil for 96-well plates.

3   Methods

3.1   Tissue Preparation

1. Treat PET membrane slides under a UV light source at 254 nm for 30 min (*see* **Note 4**).

2. Section snap-frozen tissue blocks at 5 μm and mount onto PET membrane slides. This requires an experienced operator and we recommend use of a tissue sectioning facility.

3. Air-dry at room temperature for minimum 2 h (can be up to 8 h).

4. Fix in acetone for 10 min at room temperature.

5. Air-dry and store at −80 °C.

3.2 Tissue Staining

1. Remove slides from the −80 °C freezer and thaw at room temperature.

2. Rehydrate tissues in an ethanol series for 5 min each (100, 85, and 70 %) in Coplin jars.

3. Wash slides in 1× PBS solution for 5 min with gentle agitation at room temperature.

4. Add one droplet of mounting medium containing DAPI and cover section with coverslip (*see* **Note 5**).

5. Screen the tissue to pre-select areas for subsequent tissue sampling (*see* **Note 6**).

6. To remove the coverslip, place the slide vertically in a 50 mL tube containing 1× PBS and wait for approx. 15 min for the coverslip to slide off (*see* **Note 7**).

7. Wash in fresh 1× PBS for another 5 min by tilting the tube gently to remove mounting medium from the tissue surface.

8. Dehydrate the tissue in an ethanol series (70, 85, 100 %) and leave it in 100 % ethanol until LMPC sampling or air-dry the sample and store at −80 °C.

3.3 Microdissection into Caps of PCR Tubes

To perform laser microdissection you either need to be already familiar with the technique and the system available or alternatively to seek the assistance of an experienced collaborator. Perform the microdissection in a room physically separated from PCR amplification areas. Do not enter this area with amplified DNA. The procedure described here uses the Observer Z1 (Zeiss) laser microdissection device equipped with a RoboMover that holds the PCR tube and places it above target cells to allow cell capture upon pressure catapulting. Switch on the system and let it perform initial calibration. Visually inspect the membrane slides before microdissection for blisters as these may interfere with the procedure (*see* **Note 8**).

1. Prepare ready-to-use lysis buffer by adding 2.0 μL of 5× Lysis Buffer and 2.0 μL of 5× Lysis Enzyme to 6.0 μL of PCR-grade water for every sample to be processed (*see* **Note 9**). Mix by pipetting, short-spin to collect all liquid at the bottom of the tube and place on ice.

2. Remove the slide from 100 % ethanol and let air-dry for a few minutes (*see* **Note 10**). Load the slide into the microscope stage and re-locate areas of interest as described previously (*see* **step 5** of Sect. 3.2, **Note 6**)

3. Load 10 μL of the ready-to-use lysis buffer into the cap of a 200 μL PCR-tube and mount it directly above the sample to be harvested (*see* **Note 11**).

4. Harvest the first sample by LMPC (*see* **Note 12**), carefully recover the PCR-tube from the microscope and close the PCR-tube. Immediately centrifuge the tube to collect the sample in the bottom of the tube and place the sample on ice.

5. Harvest the remaining samples including negative controls (*see* **Note 13**). For positive controls add 9 μL of ready-to-use lysis buffer to the cap of empty PCR tubes.
   Directly proceed to DNA typing.

3.4   DNA Typing Using PowerPlex 17 ESX System (Promega)

1. Again short-spin the sample tubes and incubate the samples at 75 °C in a thermal cycler for 5 min followed by an enzyme inactivation step at 95 °C for 2 min (*see* **Note 14**).

2. Thaw the PowerPlex® ESX 5× Master Mix and the PowerPlex® 17 10× Primer Pair Mix on the bench. When the reagents are completely thawed short-spin both vials for 1 s to collect all liquid at the bottom (*see* **Note 15**). Vortex both vials at full speed for 15 s and again short-spin for 1 s. Place the vials on ice.

3. Prepare an amplification master mix by adding 4.0 μL of ESX 5× Master Mix and 2.0 μL of 10× Primer Pair Mix to 4.0 μL of PCR grade water for every sample to be processed (*see* **Notes 16** and **17**), mix the solution by pipetting, short-spin and place on ice.

4. Prepare a DNA control solution using the DNA supplied with the kit or your own and dilute it to maximum 500 pg/μL (*see* **Note 18**). Add 1.0 μL of the diluted control DNA to the tube containing 9.0 μL lysis mix.

5. Add 10.0 μL of the amplification master mix to every sample and mix by flicking (*see* **Note 19**). Short-spin the tubes and place on ice.

6. Denature the samples in a thermal cycler at 96 °C for 2 min and amplify for 30 cycles as follows: Denaturation at 94 °C for 30 s followed by an annealing step at 59 °C for 2 min and an elongation step at 72 °C for 1 min 30 s. For final elongation incubate at 60 °C for 45 min.

7. Store the sample at 4 °C overnight or at −20 °C if not forwarded to post-amplification treatment within the following 24 h.

3.5   Post-amplification Treatment

1. Completely thaw the CC5 Internal Lane Standard 500, short-spin for 1 s and vortex at full speed for 15 s. Again short-spin for 1 s to collect liquid at the bottom of the tube.

2. Thaw amplified samples on the bench, mix by flicking a few times or by pipetting and short-spin.

3. Transfer the total volume of samples to a 96-well plate (*see* **Note 20**).

4. Load an additional well with 19 µL and add 1.0 µL of Power-Plex® ESX 17 Allelic Ladder Mix.

5. Add 0.5 µL of CC5 Internal Lane Standard 500 to all samples and the ladder. Fill up the remaining empty wells with water (*see* **Note 21**).

6. Cover the 96-well plate with a sealing foil, short-spin up to 1300 rpm and place on ice.

7. Denature the samples in a thermal cycler at 95 °C for 3 min and immediately chill the samples on ice. Directly proceed with capillary electrophoresis.

3.6 Analyze the Samples on any System Capable of Capillary Electrophoresis

To operate capillary electrophoresis instruments you either need to be already familiar with the technique and the system at hand or you need to seek the assistance of an experienced collaborator (e.g., core facility).

4 Notes

1. Rinse the 100 mL measuring cylinder three times with ultra-pure water and add it to the 1 L graduated flask before filling up to 1 L. The resulting $1 \times$ PBS will be pH 7.3 and does not need further adjustment.

2. The kit comes with two sealed bags. The pre-amplification components box contains the PCR master mix, primer pairs and the control DNA. The post-amplification components contain the internal standard for capillary electrophoresis. The allelic ladder mix is shipped in an extra sealed bag. Upon arrival transfer the allelic ladder mix to the post-amplification component box and store the boxes physically separated in two different areas. It is recommended that the amplified products do not go back to the pre-amplification area. We also recommend using different freezers for storage.

3. Matrix standards are initially required to calibrate the capillary analyzer. The matrix contains all dyes used in the kit and is used to compensate for cross talk. Match the matrix with the capillary electrophoresis system you use:

   PowerPlex® Matrix Standards, 310: ABI PRISM® 310 Genetic Analyzer.

   PowerPlex® Matrix Standards, 3100/3130: ABI PRISM® 3100 and 3100-Avant Genetic Analyzers and Applied Biosystems® 3130 and 3130xl Genetic Analyzers.

4. Irradiation with UV causes the membrane to become more hydrophilic. This will result in a stronger attachment of the tissue section making it less likely to detach.

5. One droplet of mounting medium is enough to cover a section area of 22×22 mm. Check for equal distribution of the medium across the section. Avoid air bubbles as tissue areas will not be sufficiently labeled with DAPI. Use larger coverslips and volumes of mounting medium if sectioned tissues cannot be covered with a 22×22 mm coverslip.

6. When screening for candidate tissue areas in the DAPI channel keep the exposure to a minimum as elongated exposure can cause DNA degradation in single cells. Once candidate areas are identified take images of the respective region or note the coordinates in order to re-locate the area. This is important because for sampling the coverslip needs to be removed and fluorescence staining bleaches quickly and important orientation may be lost. Furthermore, identify structures within your sample that will serve as reference coordinates (e.g., structures like blood vessels). This will allow you to re-locate candidate tissue areas in case a shift occurs between identifying these areas and harvesting tissue samples. Coverslips slide off easily without any agitation. This works well with slides that were covered with a coverslip up to 3 days stored in the fridge (horizontally). For longer storage of sections, we recommend sealing the coverslip with rubber gum to avoid evaporation that would cause the mounting medium to increase its viscosity which, in turn, would make it more difficult for the coverslip to slide off.

7. If the membrane of the slide shows blisters, open them at the very top by disrupting the membrane with a laser shot into the membrane. The blisters will collapse.

8. Prepare enough ready-to-use mix to run all your samples including negative and positive controls. For example, prepare 200 µL for ten samples and four controls (two negative and two positive), thus allowing for additional solution. Bear in mind that you will need to compensate for pipetting error as well as for additional samples. The latter is due to the recovering process where you might need to discard PCR-tubes (*see* **Note 11**).

9. If the slide was stored at $-80\ °C$, thaw it and place it again in 100 % ethanol. Let it air-dry and proceed with the protocol.

10. It is crucial to place the 10 µL lysis mix in the middle of the cap (having an "inner ring" structure) so that it forms a nice droplet. This setting (where the recovering droplet can be placed close to the microdissection target) yields highest recovery rates. To avoid the droplet tilting out of place attach it to the rim of the cap. In that situation, 10 µL are not sufficient to cover the whole cap. Cells that are microdissected to PCR plastic might escape analysis as DNA may stick to the plastic.

If the droplet tilts out of place discard the tube and load a new one. Take this into consideration when preparing the ready-to-use lysis mix (*see* **Note 9**).

11. We recommend checking the PCR-cap for microdissected samples. However, some samples escape this quality control (we only forward PCR-caps containing samples) as they quickly float to the side of the droplet and cannot be detected.

12. We recommend microdissecting membrane containing no sample (e.g., membrane outside the tissue section).

13. This step can be elongated up to 2 h and may vary due to parameter such as type of tissue and thickness of tissue section.

14. Check the PowerPlex® ESX 5× Master Mix for precipitate and incubate at 37 °C for a short time in order to dissolve it.

15. The obtained amplification mix is still double concentrated. This is because the amplification master mix needs to be added to the lysed samples at a 1:1 ratio which then yields the final ready-to-use concentration.

16. Prepare enough master mix to amplify all samples, e.g., make sufficient mix for ten samples and when amplifying 8.

17. Aliquot the ready-to-use control DNA. The used aliquot can be stored in the fridge at 4 °C. Store the other aliquots at −20 °C for up to 6 months. Once thawed keep in the fridge.

18. Add the amplification master mix to the tube wall. Do not mix by pipetting as this could cause loss of template DNA.

19. When running samples on the 48-capillary 3730 DNA Analyzer, be sure to add the mix to the correct wells: Load all wells of either even-numbered or odd-numbered columns (for less than 48 reactions including all samples and one well with allelic ladder). For more than 48 reactions also load the other columns.

20. This ensures that no capillary runs dry.

21. If capillary electrophoresis cannot be done directly after the post-amplification treatment, store the samples at −20 °C and perform denaturation (**steps 6** and **7** of Sect. 3.5) again before proceeding to capillary electrophoresis.

Acknowledgement

This work was supported by EU SAFE Network of Excellence (LSHB-CT-2004-503243, EU 6th Framework Package) and the County of Styria, Austria as well as a Diabetes UK grant to KMG (ref 12/0004564). We are grateful to nPOD (http://www.jdrfnpod.org/) for supplying pancreas tissues presectioned on membrane slides for these experiments and to Dr Peter Sedlmayr, Institute of Cell Biology, Medical University of Graz for facilitating the protocol development in his laboratory.

References

1. Willcox A, Richardson SJ, Bone AJ, Foulis AK, Morgan NG (2009) Analysis of islet inflammation in human type 1 diabetes. Clin Exp Immunol 155(2):173–181

2. Campbell-Thompson ML, Atkinson MA, Butler AE, Chapman NM, Frisk G, Gianani R et al (2013) The diagnosis of insulitis in human type 1 diabetes. Diabetologia 56 (11):2541–2543

3. Dotta F, Censini S, van Halteren AG, Marselli L, Masini M, Dionisi S et al (2007) Coxsackie B4 virus infection of beta cells and natural killer cell insulitis in recent-onset type 1 diabetic patients. Proc Natl Acad Sci U S A 104 (12):5115–5120

4. Richardson SJ, Leete P, Bone AJ, Foulis AK, Morgan NG (2013) Expression of the enteroviral capsid protein VP1 in the islet cells of patients with type 1 diabetes is associated with induction of protein kinase R and downregulation of Mcl-1. Diabetologia 56(1):185–193

5. Planas R, Carrillo J, Sanchez A, de Villa MC, Nunez F, Verdaguer J et al (2010) Gene expression profiles for the human pancreas and purified islets in type 1 diabetes: new findings at clinical onset and in long-standing diabetes. Clin Exp Immunol 159(1):23–44

6. Marhfour I, Lopez XM, Lefkaditis D, Salmon I, Allagnat F, Richardson SJ et al (2012) Expression of endoplasmic reticulum stress markers in the islets of patients with type 1 diabetes. Diabetologia 55(9):2417–2420

7. Vendrame F, Pileggi A, Laughlin E, Allende G, Martin-Pagola A, Molano RD et al (2010) Recurrence of type 1 diabetes after simultaneous pancreas-kidney transplantation, despite immunosuppression, is associated with autoantibodies and pathogenic autoreactive CD4 T-cells. Diabetes 59(4):947–957

8. Vanzyl B, Planas R, Ye Y, Foulis A, de Krijger RR, Vives-Pi M et al (2010) Why are levels of maternal microchimerism higher in type 1 diabetes pancreas? Chimerism 1(2):45–50

9. Ye J, Vives-Pi M, Gillespie KM (2014) Maternal microchimerism: increased in the insulin positive compartment of type 1 diabetes pancreas but not in infiltrating immune cells or replicating islet cells. PLoS One 9(1):e86985

10. Kroneis T, Gutstein-Abo L, Kofler K, Hartmann M, Hartmann P, Alunni-Fabbroni M et al (2010) Automatic retrieval of single microchimeric cells and verification of identity by on-chip multiplex PCR. J Cell Mol Med 14 (4):954–969

11. Kroneis T, Geigl JB, El-Heliebi A, Auer M, Ulz P, Schwarzbraun T et al (2011) Combined molecular genetic and cytogenetic analysis from single cells after isothermal whole-genome amplification. Clin Chem 57 (7):1032–1041

Methods in Molecular Biology (2016) 1433: 169–177
DOI 10.1007/7651_2016_331
© Springer Science+Business Media New York 2016
Published online: 01 April 2016

The Gut Microbiome in the NOD Mouse

Jian Peng, Youjia Hu, F. Susan Wong, and Li Wen

Abstract

The microbiome (or microbiota) are an ecological community of commensal, symbiotic, and pathogenic microorganisms that outnumber the cells of the human body tenfold. These microorganisms are most abundant in the gut where they play an important role in health and disease. Alteration of the homeostasis of the gut microbiota can have beneficial or harmful consequences to health. There has recently been a major increase in studies on the association of the gut microbiome composition with disease phenotypes.

The nonobese diabetic (NOD) mouse is an excellent mouse model to study spontaneous type 1 diabetes development. We, and others, have reported that gut bacteria are critical modulators for type 1 diabetes development in genetically susceptible NOD mice.

Here we present our standard protocol for gut microbiome analysis in NOD mice that has been routinely implemented in our research laboratory. This incorporates the following steps: (1) Isolation of total DNA from gut bacteria from mouse fecal samples or intestinal contents; (2) bacterial DNA sequencing, and (3) basic data analysis.

Keywords: Gut microbiome, NOD mice, 16S sequencing

1 Introduction

The NOD mouse is an in-bred mouse strain that develops spontaneous Type 1 Diabetes (T1D) and serves as an excellent animal model to study human T1D [1]. T1D is an autoimmune disease due to a T cell (both CD4[+] and CD8[+])-mediated destruction of insulin-producing islet beta cells. In addition to genetic susceptibility factors, environmental factors are believed to be associated with the initiation of the islet destruction [2, 3]. More recently, studies have drawn attention to correlations between commensal microbiota and the development of type 1 diabetes [4].

The gut microbiome (or microbiota), is a complex ecosystem of commensal, symbiotic microorganisms [5]. They are tenfold greater in number than the total human body cells [6] and protect the host from various pathogenic infections [7]. The gut bacteria digest dietary nutrients and generate energy. A healthy microbiota

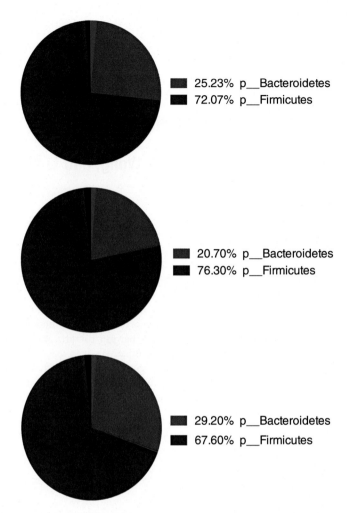

Fig. 1 Taxonomic classification of gut microbiome in female NOD mice at phylum level by age (from *top* to *bottom*: 1, 3, 5 months)

composition helps to keep the gut epithelial lining intact but maintain permeability [8, 9]. The interaction between gut epithelia and the bacteria promotes the development of a normal immune system [10, 11]. The gut microbiome has been found to be associated with the development of various diseases including obesity and type 2 diabetes [12, 13], liver disease [14, 15], intestinal inflammatory syndromes [16, 17], allergic disease [18, 19], central nervous system disease [20, 21], and autoimmune diseases [22, 23]. Among these, T1D has now been demonstrated by a number of studies including our own, to be modulated by alteration of the homeostasis of gut microbiota (Figs. 1, 2 and 3) and this can accelerate or protect from T1D development in NOD mice [24–29].

Here, we present the step-by-step protocol for gut microbiome study in NOD mice.

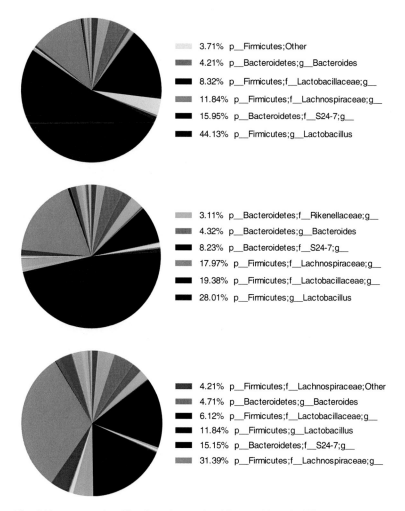

Fig. 2 Taxonomy classification of gut microbiome of female NOD mice at genus level by age (from *top* to *bottom*: 1, 3, 5 months)

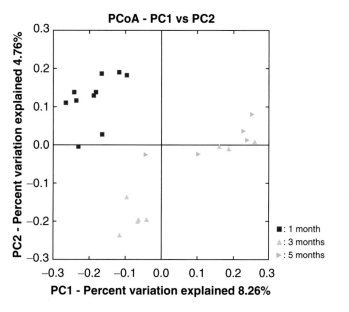

Fig. 3 Beta diversity presented as Principal Coordinate Analysis (PCoA)

2   Materials

1. TE: Tris (10 mM), bring to pH 8.0 with HCl, EDTA (1 mM).
2. 20 % SDS.
3. 20 mg/ml Proteinase K.
4. PCI: phenol–chloroform–isoamyl alcohol (25:24:1).
5. Zirconium silica beads (0.1 mm diameter) (Biospec products, Cat# 11079101z).
6. Isopropanol.
7. Ethanol (70 %)
8. Nuclease-free ddH$_2$O.
9. 1× PBS.

2.1   Equipment

1. Sterile 2 ml tube and 15 ml tube.
2. 37 °C water bath and ice bath.
3. Beads beater: miniBeadbeater-16 (Biospec products, Inc, model 607).
4. Balance.
5. 10 ml syringe.
6. 6 cm petri dish.
7. Bench top centrifuge.
8. Spectrophotometer.

3   Methods

3.1   DNA Isolation

3.1.1   DNA Isolation from NOD Fecal Sample

1. Collect fresh NOD fecal pellets into 2 ml sample tubes (*see* **Note 1**).
2. Add 300 μl TE to the tubes and vortex to loosen the pellet.
3. Add 7.5 μl SDS (20 %) to make a final concentration of 0.5 %.
4. Add 3 μl Proteinase K (20 mg/ml) to make a final concentration of 200 μg/ml.
5. Incubate the tube at 37 °C for 1 h.
6. Add 310 μl PCI and 200 μl SDS (20 %) to make a final concentration of 4.8 %, followed by a further addition of 0.3 g zirconium silica beads (0.1 mm diameter).
7. Place the tube onto the beads beater and turn on the machine for 30 s then remove the sample from the machine, keep it on ice for 1 min (*see* **Note 2**).
8. Repeat **step 7** three times, with the total time on the beads beater of 2 min.

9. Add 820 μl PCI, mix thoroughly by reversing the tube 20 times then centrifuge at $16000 \times g$ for 15 min

10. Remove aqueous layer (upper layer) to a new 1.5 ml eppendorf tube (~620–640 μl) (*see* **Note 3**).

11. Add 0.6 volume of isopropanol (~380 μl) into the tube and mix well by inverting the tube 20 times then centrifuge at $16000 \times g$ for 5 min.

12. Discard the supernatant and wash the pellet with 500 μl ethanol (70 %) then centrifuge at $16000 \times g$ for 5 min.

13. Discard the supernatant and put the tube upside down on a clean paper towel with the lid open to dry the pellet (*see* **Note 4**).

14. After the DNA is totally dried (approximately 2 h), add 100 μl ddH_2O to dissolve the DNA, by repeated pipetting followed by the measurement of the DNA concentration (*see* **Note 5**).

3.1.2   DNA Isolation from the Gut Luminal Contents of NOD Mice

1. Sacrifice the mouse and collect cecum, small intestine, large intestine into individual petri dish (kept on ice).

2. Use 5 ml sterile PBS in syringe to flush the cecum, small intestine, or large intestine, respectively, and transfer the flushed contents (*see* **Note 6**) to a new 15 ml conical tube.

3. Vortex for 2 min then centrifuge at $60 \times g$ for 3 min.

4. Transfer the aqueous layer to a new 15 ml tube and repeat **steps 2–4** twice.

5. Combine all three aqueous layers and centrifuge at $6000 \times g$ for 10 min then discard the supernatant.

6. Add 1 ml ddH_2O to resuspend the pellet followed by transferring the whole contents to a new 2 ml tube.

7. Centrifuge at $16000 \times g$ for 5 min and discard the supernatant.

8. Add 300 μl TE to the pellet and vortex to resuspend the pellet.

Go to Sect. 3.1.1, **step 3** and continue to **step 14**.

3.2   Bacterial DNA Pyrosequencing

16S ribosomal RNA (or 16S rRNA) is a component of the 30S small subunit of prokaryotic ribosomes. The genes coding for it are used in reconstructing phylogenies, due to the slow rates of evolution of this region of the gene. The common concept of gut microbiome sequencing refers to 16S rRNA sequencing.

The 16S rRNA sequencing approach enables the identification of many unculturable or unknown gut bacteria [30]. There are nine variable regions in 16S rRNA [31]. Although some primers are available [32–34] to amplify these variable regions for sequencing, the commonly used regions are V3 and V4 [35]. Table 1 shows the conservative sequences that are used to design primers for these

Table 1
Conservative regions within bacteria 16S rRNA

Location	Sequence	Region
8–27	AGAGTTTGATCCTGGCTCAG	Upstream V1
104–120	GGCGVACGGGTGAGTAA	Between V1 and V2
334–354	CCAGACTCCTACGGGAGGCAG	Between V2 and V3
519–536	CAGCMGCCGCGGTAATWC	Between V3 and V4
683–707	GTGTAGRGGTGAAATKCGYAGAKAT	Between V4 and V5
961–978	TCGATGCAACGCGAAGAA	Between V5 and V6
1053–1068	GCATGGCYGYCGTCAG	Between V6 and V7
1177–1197	GGAAGGYGGGGAYGACGTCAA	Between V7 and V8
1391–1407	TGTACACACCGCCCGTC	Between V8 and V9
1478–1492	AAGTCGTAACAAGGT	Downstream V9

Location corresponds to *E.coli* 16S rRNA nucleotide sequence location
Y: C or T; R: A or G; W: A or T; K: G or T; M: A or C; S: C or G; V: A or C or G; H: A or C or T; B: C or G or T; D: A or G or T
If amplifying V1, use "Upstream V1" and reverse complement of "Between V1 and V2" as primers, if amplifying V1–V2, use "Upstream V1" and reverse complement of "Between V2 and V3" as primers, etc.

hyper-variable regions amplification. For more information on the structure of 16S rRNA and its sequence database, please go to RDP (https://rdp.cme.msu.edu/) and/or Greengene (http://greengenes.lbl.gov/).

For 16S rRNA sequencing, the primers for the variable regions to be amplified have to be attached with different barcodes before it can be pooled and sequenced as described by Peng et al. [28].

Recent studies report the emerging use of the shotgun sequence approach to obtain the more genomic information from the gut microbiome [36]. In this case, extracted DNA can be used directly for sequencing. Shotgun sequencing provides more reads, and therefore better bacteria classification. However, the cost is higher and the data analysis requires a more advanced server platform.

3.3 Sequence Analysis

After sequencing, raw data will be generated and they need to be interpreted by specific software. The purpose of data analysis is to demultiplex sequences according to barcodes, proceeding through the sequence reads and to align with the database (such as RDP or Greengene) to find the bacterial composition. All the results are represented as relative amounts of the total content, because the total amount of the gut microbiome is still unknown.

Usually, the sequencing data are assigned operational taxonomic units (OTUs). After quality filtering based on the characteristics of each sequence, any low quality or ambiguous reads will be removed. Taxonomy assignment will be performed at various levels using representative sequences of each OTU. The most common analysis includes alpha-diversity, beta-diversity, taxonomy classification to illustrate the diversity within one group, the diversity between several groups, and individual bacterial abundance at phylum, class, order, family, genus, species level, respectively. Alpha-diversity represents the richness of gut microbiota and beta-diversity is calculated to compare differences between different microbial communities, which is shown as principal coordinate analysis (PCoA). The sequence analysis is usually done by a contracted service provider, who has specific software and powerful computer servers.

3.4 Representive Gut Microbiome in NOD Mice

Figures 1, 2 and 3 are the representative taxonomic distribution and beta-diversity results from our recent gut microbiome studies (unpublished data) in female NOD mice at different ages. Note that the gut microbiome is affected by the environment in different animal facilities, and the diabetes incidence also changes from site to site. Therefore, it is possible that our data may be different from the data generated from a different NOD colony housed in a different institution. It is also most likely that the gut microbiome in NOD mice is affected by age and the progression to T1D (the figures presented earlier were from nondiabetic mice).

4 Notes

1. Usually one fecal pellet is sufficient. If there are several, avoid collecting the particularly desiccated samples. Use a 2 ml tube with a screw cap.

2. The beads beater brings up the temperature quickly. So every 30 s of running requires cooling on ice for 1 min. If a longer time desired, repeat the step an additional cycle.

3. Be careful when removing the aqueous layer to avoid touching the interface or any of the organic layers. Trace amounts of phenol have detrimental effects on subsequent sequencing.

4. The pellet needs to be totally dry without any trace of ethanol, otherwise it will affect subsequent PCR and sequencing.

5. This can be performed on a NanoDrop or similar spectrophotometer. If the ratio of OD_{260}/OD_{280} is lower than 1.8, added equal volume of PCI to the DNA and repeat **steps 9–14** until the ratio of OD_{260}/OD_{280} is between 1.8 and 2.0 before proceeding to DNA sequencing.

6. The flushed gut contents can be frozen and stored long term at $-20\ ^\circ C$.

References

1. Makino S et al (1980) Breeding of a non-obese, diabetic strain of mice. Jikken dobutsu. Exp Anim 29(1):1–13

2. Bluestone JA, Herold K, Eisenbarth G (2010) Genetics, pathogenesis and clinical interventions in type 1 diabetes. Nature 464 (7293):1293–1300

3. Schulte BM et al (2012) Cytokine and chemokine production by human pancreatic islets upon enterovirus infection. Diabetes 61 (8):2030–2036

4. Wen L et al (2008) Innate immunity and intestinal microbiota in the development of Type 1 diabetes. Nature 455(7216):1109–1113

5. Wikipedia. Accessed from https://en.wikipedia.org/wiki/Microbiota.

6. Macpherson AJ, Uhr T (2004) Compartmentalization of the mucosal immune responses to commensal intestinal bacteria. Ann N Y Acad Sci 1029:36–43

7. Stecher B, Hardt WD (2011) Mechanisms controlling pathogen colonization of the gut. Curr Opin Microbiol 14(1):82–91

8. Natividad JM et al (2012) Commensal and probiotic bacteria influence intestinal barrier function and susceptibility to colitis in Nod1-/-; Nod2-/- mice. Inflamm Bowel Dis 18(8):1434–1446

9. Kim YS, Ho SB (2010) Intestinal goblet cells and mucins in health and disease: recent insights and progress. Curr Gastroenterol Rep 12(5):319–330

10. Dave M et al (2012) The human gut microbiome: current knowledge, challenges, and future directions. Transl Res 160(4):246–257

11. Dominguez-Bello MG et al (2011) Development of the human gastrointestinal microbiota and insights from high-throughput sequencing. Gastroenterology 140(6):1713–1719

12. Allin KH, Nielsen T, Pedersen O (2015) Mechanisms in endocrinology: Gut microbiota in patients with type 2 diabetes mellitus. Eur J Endocrinol 172(4):R167–R177

13. Tai N, Wong FS, Wen L (2015) The role of gut microbiota in the development of type 1, type 2 diabetes mellitus and obesity. Rev Endocr Metab Disord 16(1):55–65

14. Llorente C, Schnabl B (2015) The gut microbiota and liver disease. Cell Mol Gastroenterol Hepatol 1(3):275–284

15. Mafra D, Fouque D (2015) Gut microbiota and inflammation in chronic kidney disease patients. Clin Kidney J 8(3):332–334

16. Goto Y, Kurashima Y, Kiyono H (2015) The gut microbiota and inflammatory bowel disease. Curr Opin Rheumatol 27(4):388–396

17. Wright EK et al (2015) Recent advances in characterizing the gastrointestinal microbiome in Crohn's disease: a systematic review. Inflamm Bowel Dis 21(6):1219–1228

18. McCoy KD, Koller Y (2015) New developments providing mechanistic insight into the impact of the microbiota on allergic disease. Clin Immunol 159(2):170–176

19. Vital M et al (2015) Alterations of the murine gut microbiome with age and allergic airway disease. J Immunol Res 2015:892568

20. Dobbs SM et al (2015) Peripheral aetiopathogenic drivers and mediators of Parkinson's disease and co-morbidities: role of gastrointestinal microbiota. J Neurovirol 22 (1):22–32

21. Zhao Y, Lukiw WJ (2015) Microbiome-generated amyloid and potential impact on amyloidogenesis in Alzheimer's disease (AD). J Nat Sci 1(7):e138

22. Lerner A, Matthias T (2015) Rheumatoid arthritis-celiac disease relationship: Joints get that gut feeling. Autoimmun Rev 14 (11):1038–1047

23. Johnson BM et al (2015) Impact of dietary deviation on disease progression and gut microbiome composition in lupus-prone SNF1 mice. Clin Exp Immunol 181 (2):323–337

24. Alkanani AK et al (2015) Alterations in intestinal microbiota correlate with susceptibility to type 1 diabetes. Diabetes 64(10):3510–3520

25. Burrows MP et al (2015) Microbiota regulates type 1 diabetes through Toll-like receptors. Proc Natl Acad Sci U S A 112(32):9973–9977

26. Gulden E, Wong FS, Wen L (2015) The gut microbiota and Type 1 diabetes. Clin Immunol 159(2):143–153

27. Hu C, Wong FS, Wen L (2015) Type 1 diabetes and gut microbiota: friend or foe? Pharmacol Res 98:9–15

28. Peng J et al (2014) Long term effect of gut microbiota transfer on diabetes development. J Autoimmun 53:85–94

29. Kostic AD et al (2015) The dynamics of the human infant gut microbiome in development and in progression toward type 1 diabetes. Cell Host Microbe 17(2):260–273

30. Clarridge JE 3rd (2004) Impact of 16S rRNA gene sequence analysis for identification of

bacteria on clinical microbiology and infectious diseases. Clin Microbiol Rev 17(4):840–862

31. Yarza P et al (2014) Uniting the classification of cultured and uncultured bacteria and archaea using 16S rRNA gene sequences. Nat Rev 12(9):635–645

32. Soergel DA et al (2012) Selection of primers for optimal taxonomic classification of environmental 16S rRNA gene sequences. ISME J 6 (7):1440–1444

33. Chakravorty S et al (2007) A detailed analysis of 16S ribosomal RNA gene segments for the diagnosis of pathogenic bacteria. J Microbiol Methods 69(2):330–339

34. Wang Y, Qian PY (2009) Conservative fragments in bacterial 16S rRNA genes and primer design for 16S ribosomal DNA amplicons in metagenomic studies. PloS One 4(10):e7401

35. Guo F et al (2013) Taxonomic precision of different hypervariable regions of 16S rRNA gene and annotation methods for functional bacterial groups in biological wastewater treatment. PloS One 8(10):e76185

36. Deusch O et al (2014) Deep Illumina-based shotgun sequencing reveals dietary effects on the structure and function of the fecal microbiome of growing kittens. PloS One 9(7): e101021

Methods in Molecular Biology (2016) 1433: 179–207
DOI 10.1007/7651_2016_339
© Springer Science+Business Media New York 2016
Published online: 01 April 2016

Molecular Methods and Protein Synthesis for Definition of Autoantibody Epitopes

Karen T. Elvers and Alistair J.K. Williams

Abstract

Epitope mapping is the process of experimentally identifying the binding sites, or "epitopes," of antibodies on their target antigens. Understanding the antibody–epitope interaction provides a basis for the rational design of potential preventative vaccines. Islet autoantibodies are currently the best available biomarkers for predicting future type 1 diabetes. These include autoantibodies to the islet beta cell proteins, insulin and the tyrosine phosphatase islet antigen-2 (IA-2) which selectively bind to a small number of dominant epitopes associated with increased risk of disease progression. The major epitope regions of insulin and IA-2 autoantibodies have been identified, but need to be mapped more precisely. In order to characterize these epitopes more accurately, this article describes the methods of cloning and mutagenesis of insulin and IA-2 and subsequent purification of the proteins that can be tested in displacement analysis and used to monitor immune responses, in vivo, to native and mutated proteins in a humanized mouse model carrying the high-risk HLA class II susceptibility haplotype DRB1*04-DQ8.

Keywords: Epitope analysis, Mutagenesis, Protein expression, FPLC, Reverse phase HPLC, Protein purification, Mass spectrometry, Circular dichroism

1 Introduction

An understanding of the molecular basis of immune recognition of autoantigens in human autoimmune disease is essential for the development of antigen-specific immunotherapies. Antibodies to native protein antigens are often conformationally specific. Epitopes are parts of an antigen that are involved in its recognition by antibodies, B- or T-cells. The conformational epitope recognized is composed of amino acids that are separate from each other in the primary structure of the polypeptide chain but are brought together on the surface during folding. A structural epitope is a three-dimensional structure defining an antigen–antibody complex and includes all atoms of the antigen that are within predefined interatomic distances of antibody atoms. Epitope mapping is the process of locating the binding sites.

Islet autoantibodies are currently the best available biomarkers to predict future type 1 diabetes. These include antibodies to the islet beta cell proteins insulin and the tyrosine phosphatase IA-2 [1, 2]. Defining the epitopes for these antibodies could inform future therapies for type 1 diabetes.

Competitive displacement studies with animal and commercially available insulin preparations have shown that insulin autoantibodies only recognize the correctly folded molecule and are directed predominantly to residues in the A-chain loop of insulin (aa 8-13) and the N terminal of the B chain (aa 1-3) [3–5], but other amino acids may be important [6]. IAA in children who do not progress to multiple autoantibodies were shown to recognize epitopes in the C terminal of the B chain and do not recognize proinsulin [3].

Studies of truncated and chimeric protein constructs have shown that the major diabetes-associated epitope regions of IA-2A are located in the intracellular (ic) portion of the protein, with two in the juxtamembrane (JM) domain and two in the protein tyrosine phosphatase (PTP) domain [7–9]. The majority of autoantibodies to the PTP domain have been shown to recognize two epitope regions; approximately 90 % of binding to these regions in patients with T1D is abolished by modifying just two amino acids [10].

Several techniques have been developed to identify autoantibody epitope regions, including the construction of protein chimeras or deletion constructs [7, 11], competitive inhibition using peptides, animal or mutant proteins [3, 4, 9], phage display [6], and inhibition using monoclonal antibodies or Fab fragments [12, 13].

Epitopes can also be mapped using a mutagenesis approach utilizing a set of clones, each containing a unique amino acid mutation suitable for enzyme cleavage. These are then expressed in *E. coli* or the yeast *Pichia pastoris* prior to purification of the mutated proteins. For insulin, this could involve mutating various amino acids to lysine which after cleavage with Endoproteinase Lys-C, form various truncated insulins (Fig. 1).

A common approach is to study the effect of changing specific amino acids by using animal insulins or by sequential mutation of

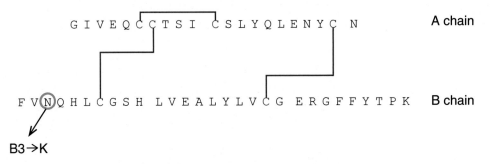

Fig. 1 DesB30 insulin (no threonine at amino acid 30 of the B chain) schematic showing an example of an amino acid mutation, B-chain asparagine to lysine. Cleavage at this lysine would give DesB1-3, 30 Insulin, a 3-amino acid truncation at the N-terminus of the B chain

IA2 (647-979)

```
G P P E P S R V S S   V S S Q F S D A A Q   A S P S S H S S T P   S W C E E P A Q A N

M D I S T G H M I L   A Y M E D H L R N R   D R L A K E W Q A L   C A Y Q A E P N T C

A T A Q G E G N I K   K N R H P D F L P Y   D H A R I K L K V E   S S P S R S D Y I N

A S P I I E H D P R   M P A Y I A T Q G P   L S H T I A D F W Q   M V W E S G C T V I

V M L T P L V E D G   V K Q C D R Y W P D   E G A S L Y H V Y E   V N L V S E H I W C

E D F L V R S F Y L   K N V Q T Q E T R T   L T Q F H F L S W P   A E G T P A S T R P

L L D F R R K V N K   C Y R G R S C P I I   V H C S D G A G R T   G T Y I L I D M V L

N R M A K G V K E I   D I A A T L E H V R   D Q R P G L V R S K   D Q F E F A L T A V

A E E V N A I L K A   L P Q
```

909→S

Fig. 2 IA-2 amino acid sequence showing mutation of cysteine to serine

single residues [14–16]; methods that have been employed for mapping insulin [3, 4] and IA-2 (Fig. 2; 10, 12, 16) autoantibody epitopes.

Effects of the changes can be assessed by competitive radio-binding assays [3, 4] and/or immunization into mice to evaluate T- and B-cell immune responses elicited by the mutant human insulin and IA-2s in mice expressing the human high risk HLA class II susceptibility haplotype DRB1*04-DQ8 [17].

This chapter describes the methods of cloning and mutagenesis of insulin and IA-2 and subsequent purification of the proteins by various fast protein liquid chromatography (FPLC) and high performance liquid chromatography (HPLC) methods to facilitate identification of key antibody epitopes.

2 Materials

1. **LB (Luria–Bertani) broth and agar:** Dissolve 10 g tryptone, 10 g NaCl, and 5 g yeast extract in 950 ml H_2O and shake until the solutes have dissolved. Adjust the final volume of the solution to 1 l with water. Sterilize by autoclaving for 20 min at 15 psi (1.05 kg/cm^2) on liquid cycle. For plates add 15 g/l agar and then autoclave. Before pouring the plates, cool agar to 55 °C before adding appropriate antibiotics, and mix the medium by swirling to avoid producing air bubbles.

   For making the low salt LB to be used with the pPICZα vector (Invitrogen) reduce the salt to 5 g/l and autoclave as above.

LB broth for glycerol stocks contains 15 % (v/v) glycerol, i.e., 15 ml glycerol in total 100 ml dH_2O.

2. **Chloramphenicol 34 mg/ml stock solution:** Dissolve 0.34 g of chloramphenicol into 10 ml 100 % ethanol. Filter through a 0.22 μm filter to sterilize. Aliquot and store at −20 °C. Use at 1:1000 dilution in LB broth or agar.

3. **Kanamycin 50 mg/ml solution:** Dissolve 0.5 g of kanamycin into 10 ml of ddH_2O. Filter through a 0.22 μm filter to sterilize. Aliquot and store at −20 °C. Use at 1:1000 dilution in LB broth or agar. For use with Rosetta cells make a 15 mg/ml stock (0.15 g in 10 ml) and use at 1:1000.

4. **TAE buffer (50×):** To make a concentrated (50×) stock solution of TAE weigh out 212 g Tris base and dissolve in approximately 375 ml deionized H_2O. Carefully add 28.55 ml glacial acid and 50 ml of 0.5 M EDTA (pH 8.0) and adjust the solution to a final volume of 500 ml. Sterilize by autoclaving. The pH of this buffer is not adjusted and should be approximately 8.5. To make the working solution of 1× TAE buffer simply dilute the stock solution by 50× in deionized H_2O.

5. **Yeast Extract Peptone Dextrose Medium (YPD):** Dissolve 10 g yeast extract and 20 g of peptone in 900 ml of H_2O. Add 20 g of agar if making YPD plates. Autoclave for 20 min on a liquid cycle. Add 100 ml of 10× D (see point 6).

6. **20 % Dextrose (10×D):** Dissolve 200 g of D-glucose in 1 l of H_2O. Autoclave or filter-sterilize.

7. **BMMY and BMGY (Buffered Glycerol Complex Medium and Buffered Methanol Complex Medium):** Dissolve 5 g yeast extract and 10 g peptone in 350 ml H_2O for BMGY (400 ml for BMMY). Autoclave for 20 min on the liquid cycle. Cool to room temperature, then add the following: 50 ml 1 M potassium phosphate buffer pH 6.0, 50 ml 10× YNB, 1 ml 500× B, and 50 ml 10× GY (for BMMY, no 10× GY is added, but 3 ml 100 % methanol instead).

8. **Yeast Nitrogen Base with Ammonium Sulfate without amino acids (10× YNB):** Dissolve 17 g of YNB without ammonium sulfate and amino acids and 50 g ammonium sulfate in 500 ml of H_2O and filter-sterilize. Store at 4 °C.

9. **0.02 % Biotin (500× B):** Dissolve 20 mg biotin in 100 ml of water and filter-sterilize. Store at 4 °C.

10. **10 % Glycerol (10× GY):** Mix 50 ml glycerol in 450 ml of water. Sterilize by autoclaving.

11. **1 M Potassium Phosphate Buffer pH 6.0:** Combine 11.5 g of K_2HPO_4 and 59.1 g KH_2PO_4 in 500 ml of H_2O and adjust the pH to 6.0 with potassium hydroxide.

12. **0.5 M EDTA solution pH 8.0:** Stir 93.05 g of disodium ethylenediamine tetraacetate $2H_2O$ into 400 ml of distilled water. Add NaOH solution to adjust the pH to 8.0 or use solid NaOH pellets. Add the NaOH solution or pellets slowly. The EDTA will slowly go into solution as the pH of the solution nears 8.0. Make up the volume to 500 ml with distilled water. Sterilize by autoclaving.

13. **Phenylmethylsulfonyl fluoride (PMSF) stock solution:** PMSF is used as an irreversible inhibitor of serine protease activity for example in protein purification. To prepare a 100 mM stock solution safely, dissolve 0.0175 g/ml PMSF in isopropanol (*see* **Note 1**). It is stable for months at 4 °C, but is hydrolyzed rapidly by water. Use at a final concentration of 1 mM.

14. **Binding buffer A for GSTrap and equilibration wash buffer for anti-His affinity:** 50 mM Tris–HCl pH 7.5, 100 mM NaCl, 1 mM DTT. Dissolve 3.765 g Trizma® Pre-set crystals (Sigma) into 500 ml Endotoxin free water (Fisher). Add 4.383 g of NaCl and 0.077 g DTT.

15. **High Salt wash buffer B for GSTrap:** 50 mM Tris–HCl pH 7.5, 300 mM NaCl, 1 mM DTT. Dissolve 3.765 g Trizma® Pre-set crystals (Sigma) into 500 ml Endotoxin free water (Fisher). Add 8.766 g of NaCl and 0.077 g DTT.

16. **Elution buffer C for GST tag:** 50 mM Tris–HCl pH 8, 10 mM reduced glutathione. Dissolve 0.606 g Tris and 0.307 g reduced glutathione in 100 ml Endotoxin free water and adjust to pH 8.

17. **Cleavage buffer for PreScission protease:** 50 mM Tris–HCl pH 7.5, 100 mM NaCl, 1 mM DTT. Dissolve 3.765 g Trizma® Pre-set crystals (Sigma) into 500 ml Endotoxin free water (Fisher). Add 4.383 g of NaCl, 0.077 g DTT, and 1 ml 0.5 M EDTA.

18. **Alkaline elution buffer for anti-His affinity:** 0.1 M Tris, 0.5 M NaCl, pH 12.0. Dissolve 1.212 g Tris and 7.305 g NaCl in 100 ml Endotoxin free water (Fisher) and adjust to pH 12.0 with NaOH.

19. **Alkaline neutralization buffer:** 1 M HCl.

20. **2× Sample loading buffer (reducing):** Add 0.5 ml 1 M Tris–HCl pH 7, 2.5 ml 20 % SDS, 2 ml glycerol, and 2 mg bromophenol blue. Make up to 10 ml with dH_2O. Mix to dissolve. To 950 μl of this add 50 μl β-mercaptoethanol.

21. **Resolving gel buffer 1.5 M Tris–HCl, pH 8.8:** For 500 ml, dissolve 90.8 g Tris base in 400 ml ddH_2O, adjust pH to 8.8 with concentrated HCl and bring the volume to 500 ml with ddH_2O.

22. **Stacking gel buffer 0.5 M Tris–HCl, pH 6.8:** For 500 ml, dissolve 30.3 g Tris base in 400 ml ddH2O, adjust pH to 6.8 with concentrated HCl and bring the volume to 500 ml with ddH$_2$O.

23. **10× Tris–glycine running buffer:** For 1 l, dissolve 30.2 g Tris base, 144 g glycine in 800 ml ddH$_2$O, add 10 g SDS and bring the volume to 1 l. Dilute to 1× working buffer.

24. **10× Transfer buffer (Western):** For 500 ml, dissolve 15.1 g Tris base, 72 g glycine in 400 ml ddH$_2$O and bring up to 500 ml. For 1 l of working buffer, add 100 ml 10× transfer buffer, 100 ml methanol, and 800 ml ddH$_2$O.

25. **10× TBST (Western wash and block buffer):** For 500 ml, dissolve 12.1 g Tris and 40 g NaCl in ddH$_2$O and adjust to pH 7.6. For 500 ml of working buffer, add 50 ml 10× transfer buffer to 450 ml ddH$_2$O and add 0.5 ml Tween 20.

26. **Destain solution (for Coomassie):** For 500 ml, mix 150 ml methanol, 50 ml glacial acetic acid, and 300 ml ddH$_2$O.

27. **Zeocin:** Commercially available. Final concentration at 25–50 μg/ml for selection of clones in LB or YPD agar.

28. **Isopropyl-beta-D-thiogalactoside (IPTG)—1 M (1000×) Stock Solution:** Weigh 2.38 g of IPTG and make up to 10 ml with sterile H$_2$O. Dissolve completely. Filter-sterilize through a 0.22 μm syringe filter. Store in 1 ml aliquots at −20 °C for up to 1 year.

3   Methods

3.1   Cloning the Gene of Interest into an Expression Vector

Choose an appropriate expression vector system (*see* **Note 2**). Design primers (*see* **Note 3**) in order that the open reading frame (ORF) of the gene of interest is cloned in frame and downstream of the promoter signal sequence and in frame with the C-terminal tag using appropriate restriction enzymes within the multiple cloning sites. If the tag is not required ensure there is a stop codon (*see* **Note 4**).

3.1.1   For Expression of Insulin in the Yeast Pichia pastoris

Proinsulin expresses poorly in yeast, but an insulin precursor (IP) containing a mini C-peptide is secreted at high levels into the supernatant [18]. Human IPs were therefore cloned into pPICZα (Invitrogen) and expressed in *P. pastoris* GS115.

3.1.2   For Expression of Islet Autoantigen-2 (IA-2) in E. coli

The intracellular portion of IA2 (647-979) was cloned from pSP64 (kind gift of Dr Vito Lampasona) into pET49b(+) (Novagen, Merck Millipore) and expressed in Rosetta™(DE3)pLysS (Novagen).

3.1.3 For Both

In intermediary cloning steps for both genes, *E. coli* library efficiency DH5α™ (competent cells) can be used.

3.1.4 General Cloning Methods

1. Amplify the gene of interest by PCR with appropriately designed primers. Prepare the PCR reaction mix on ice with 10 ng of the template DNA, 1.25 μl each of the 10 μM stock solution of the forward and reverse primers, 5 μl of 5× PCR buffer, 1 μl of a dNTP mixture (10 mM each), 1U of Phusion® High-Fidelity DNA polymerase (NEB UK or alternative), and sterile water to a final volume of 25 μl.

2. Run the PCR reaction in a thermocycler with a standard 35-cycle protocol alternating 30 s at 98 °C, 30 s at 55 °C and 1 min at 72 °C.

3. Use gel electrophoresis to determine whether the PCR reaction has been successful and whether the resulting product is of the expected size. Load 3 μl of the PCR reaction into the wells of a 1 % agarose gel in 1× TAE containing Midori Green (NIPPON Genetics Europe) with a 1 kb DNA ladder (NEB UK). Run the gel at 110 V for 35 min. The amplified product after migration can be analyzed using a Bio-Rad gel imaging system with Image Lab 5.0 software or equivalent system (Fig. 3).

4. Remove primers, nucleotides, enzymes, salts, and other impurities from the PCR product, ready for downstream processing using a QIAquick PCR Purification Kit (Qiagen UK) or alternative kit.

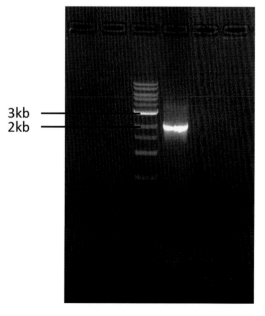

Fig. 3 Typical 1 % agarose gel stained with Midori Green. *Lane 1* shows PCR amplified gene for cloning, *Lane 2* negative control

5. Digest 1 μg of PCR product and the expression vector with the appropriate enzymes. After incubation at 37 °C, run the digested DNA on a 1 % agarose gel and extract using the QIAquick Gel Extraction Kit or equivalent.

6. The vector and digested PCR product are ligated using a 3:1 ratio of insert–vector. Typically, 100 ng of linearized plasmid is used with the corresponding amount of insert and 1 U of T4 DNA ligase (LigaFast™ Rapid DNA Ligation System, Promega) together with the ligase buffer to make the final volume to 10 μl with sterile H_2O

7. Mix 2 μl of the ligated vector and insert gently with 50 μl of thawed chemically competent *E. coli* DH5α, and incubate on ice for 30 min. Heat-shock the cells for 30 s in a 42 °C water bath and placed on ice for a further 2 min. Add SOC medium at room temperature (250 μl) to the cells and incubate at 37 °C with shaking at 225 rpm for 1 h.

8. Plate the cells on selective media; for IP and pPICZα, low salt LB with 25 μg/ml Zeocin (Invitrogen) and for IA-2 and pET49b(+), LB and 50 μg/ml kanamycin and incubate overnight at 37 °C.

9. Screen resulting colonies for insert and plasmid DNA by PCR.

3.1.5 Screening E. coli Colonies for Insert by PCR

This is a convenient high-throughput method for determining the presence or absence of insert DNA in plasmid constructs. Individual transformants are lysed in water with a short heating step and added directly to the PCR reaction.

1. Pick a few cells with sterile tip and swirl into 50 μl of dH_2O in 0.5 ml microfuge tube.

2. Heat in heating block for 10 min at 95 °C. Vortex briefly and centrifuge for 2 min at 13,000 rpm.

3. Use 1 μl as the template in a 15 μl PCR reaction.

4. Prepare the PCR reaction mix on ice: 0.75 μl each of the 10 μM stock solution of the forward and reverse primers, 1.5 μl of 10× CoralLoad PCR buffer, 0.3 μl of a dNTP mixture (10 mM each), 0.1 μl of Qiagen Taq DNA polymerase, and sterile water to a final volume of 15 μl.

5. Run the PCR reaction in a thermocycler with a standard 35-cycle protocol alternating 30 s at 94 °C, 30 s at 55 °C and 1 min at 72 °C.

6. Positive clones are determined by the presence of a PCR band of the correct size by agarose gel electrophoresis as described above (*see* **Note 5**). These clones can be inoculated into LB medium for an overnight culture and vector prepared using QIAprep Spin Miniprep Kit. Confirm the DNA sequence by automated sequencing using a Sequencing Service with appropriate primers.

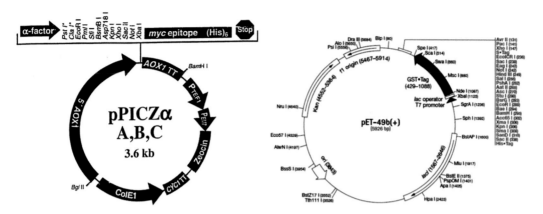

Fig. 4 The pPICZα (Invitrogen) and pET-49b(+) (Novagen) expression vectors for cloning IP and IA-2, respectively

3.2 Site Directed Mutagenesis

In vitro site-directed mutagenesis is an invaluable technique for studying protein structure–function relationships and gene expression, and for carrying out vector modification.

Make single amino acid substitutions using site-directed mutagenesis of the DNA vector (pPICZα IP and pET49b(+) IA-2, Fig. 4) using the QuikChange Site-Directed Mutagenesis kit (Agilent) or equivalent according to manufacturer's instructions. In this way a diverse collection of engineered mutant clones can be constructed to allow detailed protein structure–function analysis and help to identify specific epitopes in IP and IA-2. All mutations must be confirmed by automated DNA sequencing.

3.2.1 Method

1. Design mutagenic primers using the QuikChange Primer Design Program available online.

2. Have the primers synthesized commercially (Sigma Genosys/Aldrich or equivalent) with PAGE purification. The amount of primer to add to each reaction was calculated using the following equation.

$$X \text{ pmoles of oligo} = \frac{\text{ng of oligo}}{330 \times \text{\# of bases of oligo}} \times 1000$$

3. Prepare the DNA template using a QIAprep Spin Miniprep Kit or equivalent (as above).

4. The mutant strand synthesis reactions for thermal cycling were prepared as follows; 5 μl 10× reaction buffer, ddH₂O water up to 50 μl, 1 μl (~300 ng DNA template), 125 ng of each primer, 1 μl dNTP (10 mM), and 1 μl PfuUltra DNA polymerase.

5. Carry out the mutagenesis reaction in a thermocycler with the following parameters, 1 cycle 95 °C for 30 s; 18 cycles alternating 30 s at 95 °C, 1 min at 55 °C, and 5 min at 68 °C.

6. Place the reaction on ice to cool to ≤37 °C. Add 1 μl of *Dpn*I restriction enzyme (10 U/μl) directly to the amplification reaction and mix gently. Incubate at 37 °C for 1–2 h.

7. Precipitate the digested DNA in 2 volumes of 100 % ethanol and 1/10 volume 3 M sodium acetate. Centrifuge the solution at 13,000 rpm for 20 min to pellet the DNA, wash once with 80 % ethanol and air-dry. Resuspend the pellet in 10 μl of sterile distilled water.

8. Mix 2 μl of the mutagenesis reaction gently with 50 μl of thawed, chemically competent *E. coli* DH5α, and incubate on ice for 30 min. Heat-shock the cells for 45 s in a 42 °C water bath and place on ice for a further 2 min. Add room temperature SOC medium (500 μl) to the cells and incubate at 37 °C at 225 rpm for 1 h.

9. Plate the cells on selective media (250 μl per plate). For IP and pPICZα, low salt LB with 25 μg/ml Zeocin and for IA-2 and pET49b(+), LB and 50 μg/ml kanamycin. Incubate overnight at 37 °C.

10. Pick several colonies for plasmid DNA preparation using QIAprep Spin Miniprep Kit. Check these for the mutation by sequencing using appropriate primers.

3.2.2 Transformation into E. coli Rosetta Strains

The plasmid DNA from a clone confirmed by sequencing can be transformed into Rosetta™(DE3)pLysS (Novagen). Rosetta host strains are BL21 derivatives designed to enhance the expression of eukaryotic proteins that contain codons rarely used in *E. coli*. Gently mix 1 μl of plasmid DNA with 20 μl of cells and incubate on ice for 30 min. Heat-shock the cells for 30 s in a 42 °C water bath and placed on ice for 2 min. Add room temperature SOC medium (250 μl) to the cells and incubate at 37 °C at 225 rpm for 1 h. After incubation the cells are plated on LB with 34 μg/ml chloramphenicol and 15 μg/ml kanamycin. Individual colonies are grown up to make glycerol stocks.

3.2.3 Glycerol Stocks for Long-Term Storage of Plasmids

Bacterial glycerol stocks are important for long-term storage of plasmids. The addition of glycerol prevents damage to the cell membranes and keeps the cells viable. A glycerol stock of bacteria can be stored stably at −80 °C for many years.

1. Inoculate an overnight liquid culture in 5 ml LB and incubate at 37 °C.

2. Centrifuge to pellet the cells and resuspend in LB with 15 % glycerol. Mix gently and transfer to a screw top tube. Freeze the glycerol stock tube at −80 °C. The stock is now stable for years at −80 °C.

3. To recover bacteria from the glycerol stock, use a sterile pipette tip to scrape some of the frozen bacteria off of the top (*see* **Note 6**). Streak the bacteria onto an LB agar plate and incubate overnight. Single colonies can then be used to inoculate liquid cultures.

3.2.4 Transformation of IP into Pichia pastoris by Electroporation

Linearizing the pPICZα Construct

1. Linearize a plasmid DNA pPICZα + IP construct (~5–10 μg) by digestion within the 5′ AOX1 region using *SacI*. Precipitate the digested DNA using 2 volumes of 100 % ethanol and 1/10 volume 3 M sodium acetate.

2. Centrifuge the solution at 13,000 rpm for 20 min to pellet the DNA, wash once with 80 % ethanol and air-dry. Resuspend the pellet in 10 μl of sterile distilled water. Use 1 μl of the resuspended pellet to check by agarose gel electrophoresis for complete linearization.

Preparation of Pichia for Electroporation

1. Inoculate *P. pastoris* strain GS115 into 50 ml of YPD medium and incubate this preculture overnight at 30 °C in an incubator shaken at 250 rpm.

2. Use this culture (0.5 ml) to inoculate 500 ml of fresh YPD medium and grow overnight again at 30 °C.

3. Harvest two 50 ml aliquots of cells by centrifugation in sterile tubes at $2000 \times g$ at 4 °C for 5 min.

4. Resuspend each of the cell pellets in 50 ml ice-cold sterile water and centrifuge again.

5. Discard the supernatant and resuspend the cell pellets again in 50 ml ice-cold sterile water and centrifuge again.

6. Pool the cells and resuspend in 20 ml ice-cold 1 M sorbitol.

7. After a final centrifugation resuspend the pellet in 1 ml ice-cold sorbitol.

Transformation by Electroporation

1. Place a 2 mm gap electroporation cuvette (Bio-Rad) on ice at least 10–15 min before performing the transformation.

2. Gently mix 80 μl of competent cells (as above) with 9 μl of the linearized DNA in the microfuge tube and transfer to the chilled cuvette and incubate for 5 min on ice.

3. Pulse the cells at 1.5 V, 25 μF capacitance, and 200 Ω resistance (Bio-Rad Micropulsar). Add 1 ml of ice -cold 1 M sorbitol to the cuvette and transfer the contents to a sterile 15 ml tube.

4. Incubate the tube statically at 30 °C for 2 h to allow the cells to recover.

5. Spread 50–200 μl of cells on labeled YPD plates with 50 μg/ml Zeocin

6. Incubate plates for 2–3 days at 30 °C until colonies form.

7. Pick 10 colonies for PCR screening.

Screening Yeast Colonies for Insert by PCR

This is a convenient high-throughput method for determining the presence or absence of insert DNA in plasmid constructs. Individual transformants are lysed with a heating step and added directly to the PCR reaction.

1. Pick a few cells with sterile tip and swirl into 50 μl of 0.02 M NaOH in 0.5 ml microfuge tube.

2. Heat in heating block for 10 min at 95 °C. Vortex briefly.

3. Use 1 μl as template in 15 μl PCR reaction.

4. Prepare the PCR reaction mix on ice: 0. 5 μl each of the 10 μM stock solution of the forward and reverse primers, 1 μl of 10× CoralLoad PCR buffer, 2 μl 5× Q-solution (*see* **Note 7**), 0.4 μl of a dNTP mixture (10 mM each), 0.1 μl of Qiagen Taq DNA polymerase, and sterile water to a final volume of 10 μl.

5. Run the PCR reaction in a thermocycler with a standard 35-cycle protocol alternating 30 s at 94 °C, 30 s at 55 °C and 1 min at 72 °C.

6. Determine positive clones by the presence of a PCR band of the correct size by agarose gel electrophoresis as described above. Positive clones at this stage are selected for large scale expression.

7. The method for glycerol stocks of yeast is similar to the method above, except the yeast are spread on YPD plates and incubated at 30 °C for 2 days. The resulting growth is resuspended in 1 ml YPD broth with 15 % glycerol pipetted and stored at −80 °C in sterile screw cap vials.

3.3 Large Scale Expression of IA-2 in E. coli Rosetta™ (DE3)pLysS

1. Inoculate a single colony of Rosetta containing the pET49b(+) IA-2 wild type into 50 ml LB containing with 34 μg/ml chloramphenicol and 15 μg/ml kanamycin. Incubate this for 16 h at 37 °C at 225 rpm.

2. Use the overnight culture to inoculate 1 l LB with chlormaphenicol and kanamycin to an OD600 nm of 0.1. Incubate this culture at 37 °C at 225 rpm until the OD600 nm reads ~0.8 (about 4 h), at this point, induce expression by addition of IPTG (1 M stock) to a final concentration of 1 mM. Incubate the culture for a further 4 h with another addition of IPTG after 2 h. Take 1 ml samples at baseline and each hour for cell pellets and OD600 nm measurement. Obtain cell pellets by centrifugation at 13,000 rpm for 1 min and store at −20 °C for analysis of expression by SDS PAGE (Fig. 5).

3. Centrifuge the whole culture after 4 h expression at 8000 rpm for 20 min at 4 °C and resuspend the cell pellet in 15 ml of 15 % glycerol and transfer to a 50 ml centrifuge tube and freeze inverted at −80 °C.

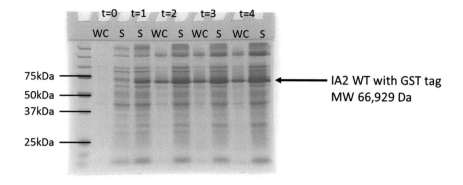

Fig. 5 Typical SDS PAGE showing time course of expression (over 4 h) of IA-2 in *E. coli* Rosetta. *WC* whole cell, *S* supernatant after lysis of samples removed during expression

3.4 Large Scale Expression of IP in P. pastoris

1. Inoculate a single colony of *P. pastoris* containing the integrated IP into 50 ml BMGY and incubate at 28 °C at 250 rpm over the weekend.

2. Inoculate 1 ml of this culture into 150 ml BMGY and incubate at 28 °C at 250 rpm overnight.

3. Pellet cells from two 50 ml aliquots from the overnight culture and resuspend both pellets in a total of 5 ml BMMY.

4. Use this cell suspension to inoculate 500 ml BMMY to OD600 nm equal to 1.0. Add one drop of sterile polypropylene glycol ($C_3H_8O_2$) to prevent foaming (*see* **Note 8**). Incubate the culture for 72 h at 28 °C at 250 rpm, feeding with 2 ml 100 % methanol at 24, 32, 48, and 56 h. Remove two 1 ml samples every 24 h, 1 ml for an OD600 nm reading and pellet 1 ml keeping the supernatant at −20 °C for analysis of expression by insulin radioimmunoassay.

5. Centrifuge the whole culture after 72 h at 8000 rpm for 20 min at 4 °C. Discard the cell pellets and filter the supernatant through a 0.45 µM filter, and add 5 ml 0.5 M EDTA and 0.5 ml 100 mM phenylmethanesulfonylfluoride (PMSF). Adjust the pH of the supernatant to 3 with 5 N HCl (~10–12 ml). Store this at 4 °C prior to SP sepharose FPLC.

3.5 Method I Purifying Expressed Protein-IA-2 from E.coli

A French press consists of a piston that is used to apply high pressure to a sample volume of 40–250 ml, forcing it through a tiny hole in the press. Efficient lysis occurs due to the high pressures and shearing used within this process.

3.5.1 French Press to Lyse Cells

1. Load the frozen cell pellet into the French Press apparatus that is prechilled to −80 °C. Pass the pellet twice through the press.

2. After lysis of the pellet resuspend in 120 ml of the following: 50 mM Tris–HCl pH 7.5, 150 mM NaCl, 1 mM DTT.

Centrifuge at 8000 rpm for 20 min at 4 °C. Transfer the supernatant to fresh tubes and centrifuge again. Finally store the supernatant in 30 ml aliquots at −20 °C for purification by FPLC.

3.6   Purification of IA-2 Using FPLC

Fast protein liquid chromatography (FPLC), was used to purify the soluble IA-2 protein from the *E. coli* cell lysates from the French Press. Separation by this method relies on the different components of the lysates having different affinities for the mobile (aqueous buffer) and stationary phase (cross-linked agarose). As the IA-2 was expressed as a fusion protein with the glutathione S-transferase (GST) tag, a GSTrap™ FF 5 ml (GE Life Sciences) column was used as the first purification step. These columns consist of a glutathione ligand coupled via a 10-carbon linker to highly cross-linked 4 % agarose, which gives high binding capacity for GST-tagged proteins. Purification using the GSTrap column was carried out on the AKTA Prime System (GE Life Sciences). This comprises an automated programmable control system, pump, fraction collector, and PrimeView software for data analysis, together with valves for buffer selection, sample injection, gradient formation, and flow diversion. The following protocol is used.

3.6.1   Loading the Cleared Cell Lysate onto the Column

1. Prepare appropriate buffers (see Materials). For loading and washing the cell lysate on the column; attach buffer A (binding buffer) and B (300 mM NaCl wash buffer) to the appropriate lines and fill the tubing and pump with these buffers using the System Wash Method.

2. Connect the GSTrap column to the system and equilibrate with binding buffer.

3. Place the tubing from buffer A into the cell lysate and load the sample onto the GSTrap column at a flow rate of 0.8 ml/min. When all the lysate is loaded, pause the system and replace the tubing into buffer A and then continue to load the lysate remaining in the tubing. Wash the column with buffer B at the same flow rate for 10 column volumes until the absorbance returns to baseline. The effluent passes through two detectors which measure salt concentration (by conductivity) and protein concentration (by absorption of ultraviolet light at a wavelength of 280 nm). This process is monitored on PrimeView software and connecting PC (Fig. 6). Collect the effluent for later analysis by SDS PAGE.

4. Remove the column from the machine. At this stage the IA-2 GST tag fusion protein is bound to the GSTrap column. It is necessary to cleave the tag from the IA-2 protein.

1. Prepare 5 ml of PreScission cleavage buffer by adding 0.5 M EDTA to buffer A at a final concentration of 1 mM.

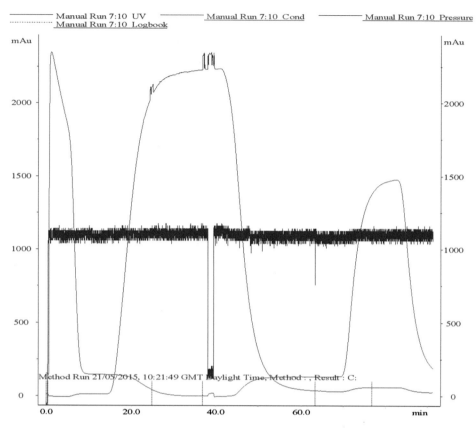

Fig. 6 Typical chromatogram loading and washing protein on to the GSTrap column. The *blue line* is the UV signal and the *red line* conductivity

3.6.2 PreScission Protease on Column Cleavage of GST-Tagged Protein Bound to GSTrap FF

2. Prepare a PreScission protease mix by adding 50 μl of PreScission Protease to 4.6 ml of cleavage buffer prepared in **step 1**.

3. Load this slowly onto the column using a syringe and Luer connector.

4. Seal the top and bottom of the column with stop plugs and incubate at 4 °C overnight.

3.6.3 Elution of the Cleaved Protein

1. Fill the machine lines with buffer A (binding buffer) and buffer C (elution buffer 50 mM Tris–HCl, 10 mM reduced glutathione, pH 8.0).

2. Use a manual run, with a flow rate of 1 ml/min and fraction collector set to 1 ml fractions.

3. Connect the column to the system and start the run.

4. The IA-2 protein is eluted first as it is no longer bound to the glutathione and this appears as a "peak" in protein concentration on the UV trace marked off in fractions (Fig. 7).

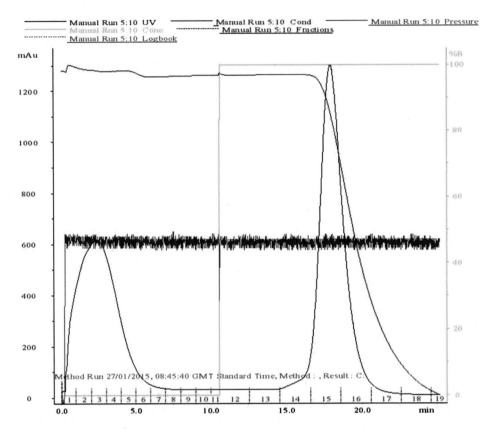

Fig. 7 Typical chromatogram showing elution of IA-2 (first peak *blue trace*) and GST-tag (second peak). The *green trace* shows the switch to 100 % elution buffer

5. When the UV trace returns to baseline after eluting the first peak, manually change the system to 100 % buffer C and fraction size 2 ml and continue elution (Fig. 7).

6. A second peak is the cleaved GST-tag.

7. The fractions, load and effluent were analyzed by SDS PAGE (Fig. 8).

3.6.4 Second Purification by Anti-His Affinity Resin

As a result from the first purification, some of the GST-tag and uncleaved protein co-elutes with the IA-2 protein. In order to further purify the IA-2 protein, the IA-2 containing fractions are pooled and passed through an anti-His affinity resin (GenScript). This is possible as the site of tag cleavage also includes a His tag. The anti-His affinity resin consists of anti-His monoclonal antibody conjugated to agarose resin.

3.6.5 Affinity Resin Preparation

1. Rinse an empty column (Pierce ThermoScientific) with equilibration wash buffer.

2. Thoroughly resuspend 1 ml of resin to form a slurry and add to the column, allowing it to settle under gravity.

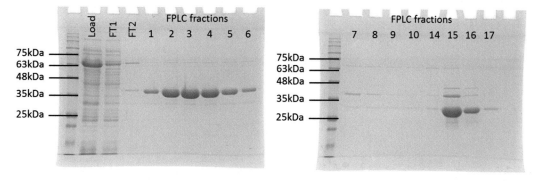

Fig. 8 SDS PAGE analysis of the fractions eluted from the GSTrap column after cleavage. FT1, effluent from column as protein is loaded. FT2, effluent whilst adding the PreScission protease. Fractions 1–10 correspond to first FPLC peak and contain cleaved IA-2. Fractions 14–17 correspond to the second FPLC peak and contain the GST tag

3. Wash the resin with 3 resin volumes of equilibration wash buffer. The column is now ready for use.

4. The column can be stored in wash buffer containing 0.02 % sodium azide at 4 °C.

3.6.6 Binding the His Tag/GST Tag Protein and Alkaline Elution

1. Load the pooled protein fractions onto the prepared column and collect the flow-through into a 15 ml tube, this contains the protein.

2. Reload the flow-through for a second time to reach maximal binding of the His tag/GST tag and maximum yield of protein.

3. Wash the resin with two 1 ml aliquots of wash buffer.

4. Use an alkaline elution procedure to elute the bound His tag/ GST tag. Label six tubes and add 50 μl of 1 M HCl for neutralization to each tube.

5. Add six 1 ml aliquots of alkaline elution buffer (0.1 M Tris, 0.5 M NaCl, pH 12.0), allowing each aliquot to drain from the column into the separately labeled tubes.

6. Finally wash the resin with wash buffer before storage (as **step 4** above).

7. Analyze the various fractions and flow-throughs by SDS PAGE (Fig. 9).

3.7 Endotoxin Removal

Endotoxins are lipopolysaccharides (LPS) derived from cell membranes of gram-negative bacteria and are responsible for its organization and stability. Although endotoxins are linked within the bacterial cell wall, they are continuously released during cell growth and division as well as at cell lysis during protein purification. Removal of endotoxin is one of the most difficult downstream processes during protein purification. The maximum level of

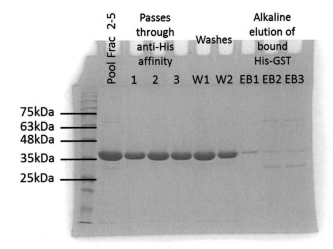

Fig. 9 Typical SDS PAGE of IA-2 after anti-His affinity removal of uncleaved IA-2 and co-eluted GST tag

endotoxin for intravenous applications in humans of pharmaceutical and biologic product is set to 5 endotoxin units (EU) per kg of body weight per hour [19]. For mice (30 g body weight) this translates to 6 EU/mg at a dose of 0.025 mg/h [20]. Many commercial products are available to remove endotoxin from recombinant proteins. The Detoxi-Gel endotoxin removing columns (Thermo Scientific) use immobilized polymyxin B to bind and remove endotoxins from proteins. Use these according to the manufacturer's protocol. The level of endotoxin in the final protein can be quantified using the Thermo Scientific Pierce LAL Chromogenic Endotoxin Quantification Kit with a limit of detection of 0.1 EU/ml. Use this according to the manufacturer's protocol.

3.8 SDS PAGE, Coomassie Blue Staining, and Western Analysis

As SDS polyacrylamide gel electrophoresis (PAGE) and Western blot analysis are widely used techniques, the methods are described briefly. For IA-2, these methods are used to show time course of protein expression, elution of protein from the GSTrap, and the removal of the GST-tag from the protein after anti-HIS affinity purification.

3.8.1 Protein Separation by Gel Electrophoresis

1. Mix the protein samples with sample loading buffer (reducing).

2. Prepare resolving and stacking gels at an adequate percentage of acrylamide based on the molecular mass of the protein of interest. For IA-2 this is a 12 % resolving gel and 6 % stacking gel (Table 1).

3. Cast the resolving gel and gently overlay with isopropanol to accelerate polymerization. After polymerization, remove the isopropanol and pour the stacking gel, gently inserting the gel comb without introducing air bubbles.

Table 1
Components of gels for SDS PAGE

16 ml of 12 % resolving gel		10 ml of 6 % stacking gel	
5.3 ml	ddH$_2$O	5.3 ml	ddH$_2$O
6.4 ml	30 % Acrylamide	2 ml	30 % Acrylamide
4 ml	1.5 M Tris pH 8.8	2.5 ml	0.5 M Tris pH 6.8
160 µl	10 % SDS	100 µl	10 % SDS
160 µl	10 % APS	100 µl	10 % APS
16 µl	TEMED	10 µl	TEMED

4. When the gel is set, load the samples and protein ladder.

5. Run the electrophoresis at 110 V for approx. 100 min until the dye front reaches the bottom of the gel in 1× Tris–glycine running buffer.

6. Following the electrophoresis, remove the gel carefully and either transfer to PhastGel Blue R stain (Sigma) or continue with Western blotting. After staining for 1 h remove from the stain and destain.

3.8.2 Transferring the Protein from the Gel to Membrane

1. Transfer the proteins from the gel to supported nitrocellulose membrane by electrophoretic transfer using a Mini Trans-Blot Cell (Bio-Rad) according to the instructions provided by the manufacturer. The gel is run at 250 mA for 60 min.

2. After electrophoretic transfer, remove the membrane from the transfer unit and block with 5 % BSA in 1× TBST for 1 h.

3. Incubate the membrane overnight with primary antibody (GST (B-14), Monoclonal Antibody 1:3000 dilution, Santa Cruz Biotechnology) in 1× TBST.

4. Following overnight incubation, remove the primary antibody and wash the membrane six times with 1× TBST (~5 min per wash).

5. Incubate the membrane with secondary antibody (anti-mouse IgG peroxidase, Sigma Aldrich) for 1 h and then wash six times with 1× TBST (~5 min per wash).

6. Develop the membrane by incubating with Clarity Western ECL chemiluminescent substrate (Bio-Rad). Detect immunoreactive bands using a Bio-Rad Chemidoc MP with Image Lab 5.0 software (Fig. 10).

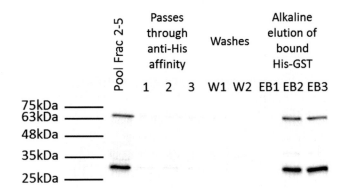

Fig. 10 Typical Western analysis of IA-2 after anti-His affinity removal of uncleaved IA-2 (*top* band) and co-eluted GST tag (*bottom* band)

3.9 Method II Purifying Expressed Protein: IP from Pichia

3.9.1 FPLC Using SP Sepharose Fast Flow Strong Cation Exchanger

Ion exchange chromatography is based on the reversible interaction between a charged protein and an oppositely charged ion exchange medium. The net surface charge of a protein varies according to the surrounding pH. If the pH is below the isoelectric point (pI), a protein will bind to a negatively charged cation exchanger. For purification of IP, the pH was adjusted below the pI 4.74, therefore SP Sepharose FF (GE LifeSciences) was used. This consists of 6 % agarose cross linked to a sulfopropyl ($-CH_2CH_2CH_2SO_3-$) group. The protein binds as it is loaded, becoming concentrated on the column. Uncharged proteins, or those with the same charge as the ionic groups, elute in the flow-through. Increasing ionic strength (using a gradient e.g., increase in pH) displaces bound proteins as ions in the buffer compete for binding sites and thus the bound substances are eluted at different times.

1. Mix the SP Sepharose slurry (50–60 ml) thoroughly and pack into a XK26/20 column (GE Life Sciences) connected to an AKTA Prime system (GE Life Sciences) according to manufacturer's protocol.

2. Prepare appropriate buffers (see Materials). For loading and washing the culture supernatant on the column, attach buffer A (50 mM acetic acid) and B (50 mM ammonium acetate adjusted to pH 7.5 with ammonium hydroxide) to the appropriate lines and fill the tubing and pump with these buffers using the System Wash Method.

3. Before each chromatographic run, equilibrate the packed column by washing with at least 150 ml of 50 mM acetic acid pH 3 (buffer A) at a flow rate of 5 ml/min or until the effluent shows stable conductivity and UV trace.

4. Place the line to buffer A in the culture supernatant that is kept on ice and load onto the column at 3 ml/min. Once loaded the line is transferred back to buffer A and continued to load and wash until the UV trace returns to baseline.

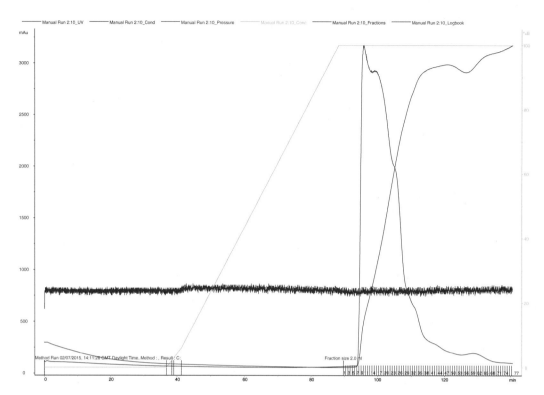

Fig. 11 Typical chromatogram showing elution of IP from SP Sepharose using a pH gradient

5. Elute the protein at 2 ml/min with a pH gradient (0–100 % buffer B) over 75 min starting fraction collection (2 ml aliquots) after 90 min (Fig. 11). Assay the peaks corresponding to eluted protein and fractions using radioimmunoassay.

6. Pool the fractions containing IP and adjust the pH back to 3.0. These pooled fractions are loaded onto 2× 1 ml SP Sepharose columns (Fig. 12).

7. The method for loading and elution from the smaller SP columns is essentially as described in **steps 1–6** above. Gradient elution at a flow rate of 1 ml/min is from 0 to 100 % buffer B over 75 min with 2 ml fraction collection (Fig. 12). Assay the fractions corresponding to the peaks by radioimmunoassay. Pool the final fractions that contain IP and freeze-dry.

3.10 Dialysis of Culture Supernatant Before Ion Exchange

Dialysis is a separation technique that removes small, unwanted compounds in solution by selective and passive diffusion through a semipermeable membrane. For IP that is secreted into the culture supernatant dialysis cleans up the solution, allowing better binding to the SP column. However, it does also precipitate a proportion of the protein which must be processed downstream by further methods. The culture supernatant (sample) is poured into a suitable dialysis membrane and sealed. It is placed in a buffer

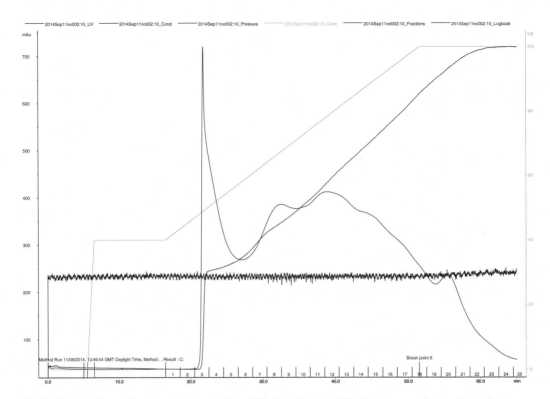

Fig. 12 Typical chromatogram showing elution of IP from 2× 1 ml SP Sepharose using a pH gradient

solution (dialysate). The protein that is larger than the membrane-pores is retained within the membrane, but small molecules and buffer salts pass freely through the membrane, reducing the concentration of those molecules in the sample. Changing the dialysate buffer allows more contaminants to diffuse into the dialysate. In this way, the concentration of small contaminants within the sample can be decreased to acceptable levels.

3.10.1 Preparation of Dialysis Tubing

Cut the Visking dialysis membrane (7000MWCO, Medicell Membranes), to size for 250 ml volume. Heat the membrane in 1.5 l of 2 % sodium bicarbonate and 1 mM EDTA at 80 °C for 30 min. Rinse the tubing in pyrogen-free distilled water.

1. Fill the dialysis tubing with culture supernatant at pH 3 and dialyze overnight in cold room in a 2 l measuring cylinder with 50 mM acetic acid pH 3 with mixing.

2. Change the acetic acid buffer before dialyzing for another 1 h at room temperature.

3. Centrifuge the supernatant at 8500 rpm for 20 min to remove any precipitated protein or particulates. Load the dialyzed supernatant onto the SP column as described above.

3.11 Insulin Radioimmunoassay

This assay was used to test which fractions from FPLC contained the IP and has been adapted from a method described in detail elsewhere [21]. Briefly, it is as follows:

1. Pipette 5 μl of HUI-018 anti-human insulin monoclonal antibody (Dako) diluted 1/2000 with normal human serum (negative pool) into the wells of a deep-well plate (except for the wells containing high and low controls)

2. Then pipette 5 μl of human insulin standards (e.g., Actrapid, diluted in TBT) spanning the range 1 U/ml to 0.2 mU/ml (and 0, TBT alone) in duplicate into the wells containing HUI-018 antibody.

3. Pipette 5 μl of supernatant undiluted, diluted 1:10 and/or 1:100 with TBT from the expression time course and the undiluted FPLC fractions into the remaining wells containing HUI-018 antibody.

4. Finally pipette 5 μl of the control sera into additional empty wells.

5. 12,000 cpm ^{125}I human insulin (>2000 Ci/mmol, Perkin Elmer, NEX420050UC) in 25 μl TBT is pipetted into each well of the plate containing antibody or control sera.

6. The plate is spun down briefly, mixed and incubated overnight in the fridge.

7. Fifty microliters of washed PGS suspension in TBT (10 μl/well) is added to each well containing reaction mixture, and the plate is spun briefly and incubated with shaking in the fridge for 1 h 45 min.

8. Wash the plate five times with 800 μl/well TBT by centrifugation at $500 \times g$ using 8-well dispenser (Nunc) and aspirator (Sigma) manifolds. After the final wash, transfer the pellet to microtubes using an 8-way multichannel pipette and count on a gamma counter for 15 min.

3.12 Reverse Phase HPLC

After the fractions has been freeze-dried further purification is carried out by reverse phase high performance liquid chromatography (HPLC). HPLC relies on protein hydrophobicity to separate proteins. The stationary phase is hydrophilic and has a strong affinity for hydrophilic molecules in the mobile phase, and thus, they bind to the column. Hydrophobic molecules pass through the column and are eluted first. The hydrophilic molecules can then be eluted from the column by increasing the polarity of the solution in the mobile phase. The length of time a protein takes to pass through the column depends on how much it interacts with the stationary phase and is termed its retention time.

| 3.12.1 For Purification of IP | *Equipment:* The HPLC system is composed of a quaternary delivery pump, degasser, thermostatic column compartment, autosample, VWD and fraction collector (Agilent 1260 Infinity). The HPLC system is equipped with two different columns: an analytical Eclipse Plus C18, 3.5 μm particle size, 4.6 mm internal diameter × 100 mm length (Agilent) and a semi-preparative Eclipse XDB C18, 5 μm particle size, 9.4 mm × 250 mm (Agilent). Equipment control, data acquisition and integration were controlled by PC with OpenLAB CDS ChemStation data handling system. |

Chromatographic conditions: The mobile phase A consists of water with 0.1 % TFA and mobile phase B acetonitrile (ACN) and 0.1 % TFA. The gradient elution profile for each column is described in Table 2.

The column temperature is maintained at 20 °C. Peak responses are measured at 215 nm using a variable wavelength detector (VWD).

| 3.12.2 Method | 1. Resuspend freeze-dried IP in 100 μl 1 mM HCl and 100 μl water/0.1% TFA/20% ACN before injection into the HPLC using the Eclipse XDB 9.4 mm column and associated conditions (Fig. 13). |

2. Collect the peaks with retention time corresponding to uncleaved IP and pool for freeze drying.

| 3.12.3 Endoproteinase Lys-C Digestion | The HPLC purified IP requires cleavage at the lysine to remove the leader sequence and the short C-peptide joining the A and B chains. Cleavage is also necessary for truncated insulins where amino acids were mutated to additional lysines. This is achieved by Endoproteinase Lys-C digestion. This enzyme from *Lysobacter* |

Table 2
Characteristics of HPLC columns and gradient elution profile

Eclipse plus 4.6 mm		Eclipse XDB 9.4 mm	
Time (min)	%B acetonitrile	Time (min)	%B acetonitrile
0	10	0	10
3.32	30	8.5	30
10	45	25	45
12	80	27	80
16	80	32	80
16.67	10	34	10
20	10	40	10
1 ml/min		4 ml/min	

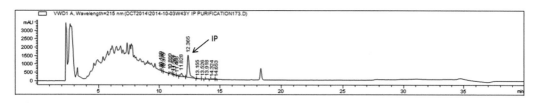

Fig. 13 Typical HPLC chromatogram showing purification of the peak corresponding to IP

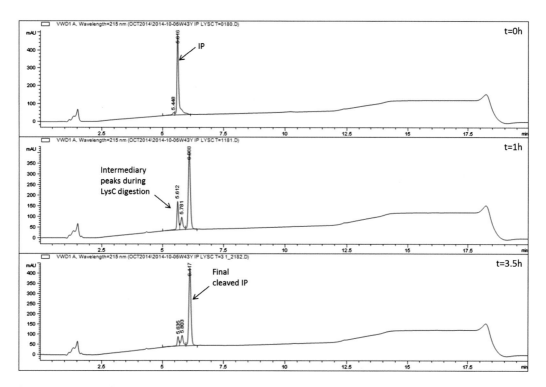

Fig. 14 Typical HPLC chromatogram showing Endoproteinase LysC digestion over time

enzymogenes is a serine endoprotease, which specifically cleaves peptide bonds at the carboxyl side of lysine.

1. Dissolve the IP in 100 μl 100 mM Tris–HCl, pH 8.5.

2. Add resuspended Endoproteinase Lys-C (0.4 μg) to the protein and mix gently.

3. Incubate this for 3–5 h at 37 °C. Monitor the digestion by HPLC using the Eclipse Plus 4.6 mm at baseline, 1, 3, and/or 5 h. Remove a 3 μl aliquot from the digest reaction and mix with 100 μl water/0.1 % TFA/10 % ACN before injection into the HPLC (Fig. 14).

4. The digest reaches completion, when the peak analysis at each time point remains unchanged. The peaks are then collected by HPLC and freeze dried.

**3.13 Mass
Spectrometry**

Confirmation of the proteins mass was carried out by the EPSRC UK National Mass Spectrometry Facility at Swansea University. The facility performs MALDI-TOF spectrometry on an Applied Biosystems Voyager DE-STR. The procedure involves mixing solutions of sample protein in acetonitrile/0.1%TFA and matrix, e.g., sinapinic acid in a 1:1 ratio, and pipetting 0.5–1 μl onto the target well of a sample plate. The sample spot is dried, allowing co-crystallization of the mixture, then irradiated with a pulsed N_2 laser (337 nm, $f = 3$ or 20 Hz). The sample is desorbed and ionized, then accelerated into a flight tube (typically 20 kV). The instrument may be run in any combination of positive or negative and linear or reflector modes. The resulting chromatogram gives the molecular weight of the protein (Fig. 15).

**3.14 Circular
Dichroism**

Circular dichroism (CD) spectroscopy was carried out in University of Bristol Chemistry Department. It is a rapid method that determines the secondary structure and folding properties of proteins that have been expressed and purified and also determines whether a mutation affects its conformation or stability. Circular dichroism is the difference in the absorption of left-handed circularly polarized light (L-CPL) and right-handed circularly polarized light (R-CPL) and occurs when proteins contain one or more light-absorbing groups. CD spectra are collected in high transparency rectangular quartz cuvettes (cells) with 1 mm pathlength. The buffers for CD spectroscopy must not contain any materials that are optically active and should be as transparent as possible, e.g., 100 mM sodium phosphate pH 7.5. Samples for CD spectroscopy must be at least

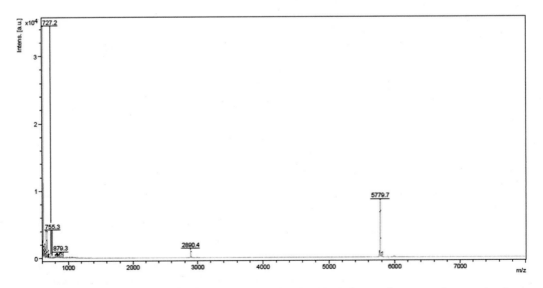

Fig. 15 Typical mass chromatogram showing mutated insulin where the *x*-axis represents mass to charge ratio and the *y*-axis represents signal intensity. The mass determined mass was 5779.7 and predicted formula weight mass 5780

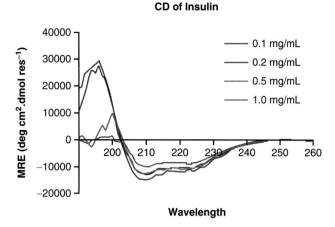

Fig. 16 Circular dichroism spectra of insulin (Sigma) at different concentrations in 100 mM sodium phosphate pH 7.5

95 % pure and at a concentration of at least 0.1 mg/ml. The protein concentration was determined using a NanoPhotometer P360 (Implen).

1. Select an appropriate wavelength range and step size, e.g., "far-UV" 260–180 nm at 0.5 nm step.

2. Select an appropriate bandwidth and time per point, e.g., 1 nm BW and 0.5 s per point.

3. Perform a baseline measurement of buffer only.

4. Replace the buffer with sample and put the cell in the cell holder taking care to maintain its orientation.

5. Acquire a spectrum of the sample. Repeat with further samples making sure the cuvette is cleaned thoroughly between samples (*see* **Note 9**).

6. Analyze the data using DichroWeb online analysis for protein Circular Dichroism spectra (Fig. 16).

4   Notes

1. PMSF is extremely toxic, and preparation of solutions should be done very carefully and in the fume hood.

2. *E. coli* or yeast are most suited to large scale production of proteins. Both can be cultured to high cell densities with minimum and low cost complexity of growth media. High levels of expression can be achieved, in *E. coli* the protein is secreted into the periplasm whilst in yeast it is secreted into the medium. Yeast has advantages in that it is eukaryotic and proteins are therefore post translationally modified and folded.

As for expression vector consider the size of the insert, copy number of the plasmid (high copy number offers greater yields), restriction sites within the multiple cloning site and the antibiotic resistance marker.

3. Considerations for primers: The 3′-end of the primer molecule is critical for the specificity and sensitivity of PCR. It is recommended not to have 3 or more G or C bases at this position. Primer pairs should be checked for complementarity at the 3′-end. This often leads to primer-dimer formation. Bases at the 5′-end of the primer are less critical for primer annealing. Therefore, it is possible to add restriction sites, to the 5′-end of the primer molecule to aid cloning adding a few bases overhang to allow efficient cleavage. Primer length of 18–30 bases is optimal for most PCR applications. The melting temperature of both primers should be similar.

4. To increase the efficiency of termination it is possible to use 2 or 3 stop codons in series.

5. The PCR product can be loaded onto agarose gels directly without addition of loading buffer because coral load was used.

6. Do not let the glycerol stock unthaw. Subsequent freeze and thaw cycles reduce shelf life and viability of the stored bacteria.

7. Q-solution is critical for this protocol; the main ingredient is betaine. It improves the amplification of DNA by reducing the formation of secondary structures.

8. When up-scaling to large shake-flasks a particular problem is foaming, which is commonly prevented by the addition of antifoam, e.g., sterile polypropylene glycol. Foaming can lead to reduced yields since bursting bubbles can damage proteins and can also result in a loss of sterility if the foam escapes.

9. Clean cells are the foundation of any spectrophotometric analysis. The residue from previous analysis will cause inaccuracies, low sensitivity and lack of precision. Rinse with ethanol and copious amounts of distilled water. Also inspect the condition of the cells. If they are cracked, chipped or scratched it is important to replace the cells with new ones.

References

1. Palmer JP (1987) Insulin autoantibodies—their role in the pathogenesis of IDDM. Diabetes Metabol Rev 3(4):1005–1015

2. Notkins AL, Lan MS, Leslie RDG (1998) IA-2 and IA-2 beta: the immune response in IDDM. Diabetes Metabol Rev 14(1):85–93

3. Achenbach P, Koczwara K, Knopff A, Naserke H, Ziegler AG, Bonifacio E (2004) Mature high-affinity immune responses to (pro)insulin anticipate the autoimmune cascade that leads to type 1 diabetes. J Clin Invest 114:589–597

4. Castano L, Ziegler AG, Ziegler R, Shoelson S, Eisenbarth GS (1993) Characterization of insulin autoantibodies in relatives of patients with type I diabetes. Diabetes 42:1202–1209

5. Koczwara K, Muller D, Achenbach P, Ziegler AG, Bonifacio E (2007) Identification of insulin autoantibodies of IgA isotype that

preferentially target non-human insulin. Clin Immunol 124:77–82

6. Devendra D, Galloway TS, Horton SJ, Evenden A, Keller U, Wilkin TJ (2003) The use of phage display to distinguish insulin autoantibody (IAA) from insulin antibody (IA) idiotypes. Diabetologia 46:802–809

7. Bearzatto M, Naserke H, Piquer S, Koczwara K, Lampasona V, Williams A, Christie MR, Bingley PJ, Ziegler AG, Bonifacio E (2002) Two distinctly HLA-associated contiguous linear epitopes uniquely expressed within the islet antigen 2 molecule are major autoantibody epitopes of the diabetes-specific tyrosine phosphatase-like protein autoantigens. J Immunol 168:4202–4208

8. Zhang B, Lan MS, Notkins AL (1997) Autoantibodies to IA-2 in IDDM: location of major antigenic determinants. Diabetes 46:40–43

9. Seissler J, Schott M, Morgenthaler NG, Scherbaum WA (2000) Mapping of novel autoreactive epitopes of the diabetes-associated autoantigen IA-2. Clin Exp Immunol 122:157–163

10. Elvers KT, Geoghegan I, Shoemark DK, Lampasona V, Bingley PJ, Williams AJ (2013) The core cysteines, (C909) of islet antigen-2 and (C945) of islet antigen-2β, are crucial to autoantibody binding in type 1 diabetes. Diabetes 62:214–222

11. Bonifacio E, Lampasona V, Bernasconi L, Ziegler A-G (2000) Maturation of the humoral autoimmune response to epitopes of GAD in preclinical childhood type 1 diabetes. Diabetes 49:202–208

12. Weenink SM, Lo J, Stephenson CR, McKinney PA, Ananieva-Jordanova R, Rees Smith B, Furmaniak J, Tremble JM, Bodansky HJ, Christie MR (2009) Autoantibodies and associated T-cell responses to determinants within the 831-860 region of the autoantigen IA-2 in Type 1 diabetes. J Autoimmun 33:147–154

13. Padoa CJ, Banga JP, Madec A-M, Ziegler M, Schlosser M, Ortqvist E, Kockum I, Palmer J, Rolandsson O, Binder KA, Foote J, Luo D, Hampe CS (2003) Recombinant Fabs of human monoclonal antibodies specific to the middle epitope of GAD65 inhibit type 1 diabetes-specific GAD65Abs. Diabetes 52:2689–2695

14. Kristensen C, Kjeldsen T, Wiberg FC, Schäffer L, Hach M, Havelund S, Bass J, Steiner DF, Andersen AS (1997) Alanine scanning mutagenesis of insulin. J Biol Chem 272:12978–12983

15. Chen H, Shi M, Guo ZY, Tang YH, Qiao ZS, Liang ZH, Feng YM (2000) Four new monomeric insulins obtained by alanine scanning the dimer-forming surface of the insulin molecule. Protein Eng 13:779–782

16. McLaughlin KA, Richardson CC, Williams S, Bonifacio E, Morgan D, Feltbower RG, Powell M, Rees Smith B, Furmaniak J, Christie MR (2015) Relationships between major epitopes of the IA-2 autoantigen in type 1 diabetes: implications for determinant spreading. Clin Immunol 160:226–236

17. Wen L, Chen NY, Tang J, Sherwin R, Wong FS (2001) The regulatory role of DR4 in a spontaneous diabetes DQ8 transgenic model. J Clin Invest 107:871–880

18. Kjeldsen T, Pettersson AF, Hach M (1999) Secretory expression and characterization of insulin in Pichia pastoris. Biotechnol Appl Biochem 29(Pt 1):79–86

19. Daneshian M, Guenther A, Wendel A, Hartung T, von Aulock S (2006) In vitro pyrogen test for toxic or immunomodulatory drugs. J Immunol Methods 313(1-2):169–175

20. Malyala P, Singh M (2008) Endotoxin limits in formulations for preclinical research. J Pharm Sci 97(6):2041–2044

21. Wyatt R, Williams AJ (2015) Islet autoantibody analysis: radioimmunoassays. Methods Mol Biol. [Epub ahead of print] PMID: 26659803

Methods in Molecular Biology (2016) 1433: 209–213
DOI 10.1007/7651_2016
© Springer Science+Business Media New York 2016

INDEX